4th Computational Particle Physics Workshop (CPP 2016)

Journal of Physics: Conference Series Volume 920

Hayama, Japan
8 – 11 October 2016

ISBN: 978-1-5108-5213-6
ISSN: 1742-6588

Printed from e-media with permission by:

Curran Associates, Inc.
57 Morehouse Lane
Red Hook, NY 12571

Some format issues inherent in the e-media version may also appear in this print version.

Printed by Curran Associates, Inc. (2017)

For permission requests, please contact the Institute of Physics
at the address below.

Institute of Physics
Dirac House, Temple Back
Bristol BS1 6BE UK

Phone: 44 1 17 929 7481
Fax: 44 1 17 920 0979

techtracking@iop.org

Additional copies of this publication are available from:

Curran Associates, Inc.
57 Morehouse Lane
Red Hook, NY 12571 USA
Phone: 845-758-0400
Fax: 845-758-2633
Email: curran@proceedings.com
Web: www.proceedings.com

Table of contents

Volume 920

4th Computational Particle Physics Workshop
8–11 October 2016, Hayama, Kanagawa, Japan

Accepted papers received: 30 October 2017
Published online: 9 November 2017

Preface

Papers

IOP Conf. Series: Journal of Physics: Conf. Series **920** (2017) 011001 doi:10.1088/1742-6596/920/1/011001

Introduction

The fourth Computational Particle Physics Workshop (CPP2016) was successfully carried out in Hayama, Japan. The main purpose of this workshop was to discuss issues on theoretical analysis with computers in large scale analysis in the field of high energy physics. Since Japanese, French, and Russian collaborating research groups held the very first Workshop in Tokyo in 1998, participants from other European and Asian countries started to join.

This Workshop was hosted in honor of commemorating Professor Yoshimitsu Shimizu's dedication to this field of study. He passed away on the 6th of October, 2015. From here on, we shall call Professor Shimizu "Shimizu-sensei," the way he used to be called by other researchers and his friends, as in accordance with a Japanese custom.

Shimizu-sensei has achieved highly in the analysis of higher order perturbation effect in the electro-weak theory. Together he also found out and realized that analysis in this particular field would progress revolutionary with the advanced use of computers. At the same time, he played the center role on constructing and analyzing such system, and achieved to establish today's calculation methods. He has also established close friendship with many researchers worldwide in this field of study especially for his generous personality.

Shimizu-sensei played the main role in the past three CPP Workshops as well, wishing for further progress in this field of study and its future until the day he passed away. This Workshop was dedicated to express his passion and will toward his lifetime research. Furthermore, according to his last will and testament, a portion of his inheritance supported the operation of this Workshop. At this time, the Organizing Committee decided to hold memorial sessions inviting those who were close to him in his living days in addition to scientific sessions.

We have summarized scientific session presentations into these proceedings. We believe that they will provide both the current status of the study in this field, and the directions and suggestions for future progress.

The Committee would like to express a sincere and deepest appreciation towards the ones who contributed and paid dedication towards this Workshop, participants, and those who participated the memorial sessions.

CPP2016 Organizing Committee

IOP Conf. Series: Journal of Physics: Conf. Series **920** (2017) 011002 doi:10.1088/1742-6596/920/1/011002

Peer review statement

All papers published in this volume of *Journal of Physics: Conference Series* have been peer reviewed through processes administered by the proceedings Editors. Reviews were conducted by expert referees to the professional and scientific standards expected of a proceedings journal published by IOP Publishing.

IOP Conf. Series: Journal of Physics: Conf. Series **920** (2017) 012001 doi:10.1088/1742-6596/920/1/012001

A novel method to compute one loop 4 point functions

J.Ph. Guillet[1], E. Pilon[1] and S. Zidi[2]

[1] LAPTh, Université Savoie Mont Blanc, CNRS, B.P.110, Annecy-le-Vieux F-74941, France
[2] LPTh, Université de Jijel, B.P. 98 Oued-Aissa, 1800 Jijel, Algérie

E-mail: guillet@lapth.cnrs.fr

Abstract. We present a part of an ongoing work initiated by Prof. Shimizu on a method to compute numerically two loop scalar integrals as a sum of two dimensional integrals of generalized scalar one loop four point functions. This implies to have a formula for the one loop scalar four point functions which is valid outside the physical domain, this is the object of the present work.

1. Introduction

This work has been started after a visit of Prof. Shimizu in our institute, he explained to us his idea about computing scalar two loop 3 and 4 point functions numerically as a sum of two dimensional integrals of generalized scalar one loop four point functions. He gave us his notes (a part in English and a part in Japanese!) and he trusted us to push forward this project. I (the author, J.Ph. G.) would like to take this opportunity to thank him for introducing me to the Japanese culture (despite the fact that I was not a diligent student!) and for his kindness, I really enjoyed to work with him.

More precisely, Prof. Shimizu showed that any scalar two loop three point function can be written as

$$I^4_{3\,\text{2-loop}} = \sum \int_0^1 d\xi \int_0^1 d\rho \; \tilde{I}^4_4(\xi, \rho) \tag{1}$$

and any scalar two loop four point function can be written as

$$I^4_{4\,\text{2-loop}} = \sum \int_0^1 d\xi \int_0^1 d\rho \; \tilde{I}^4_5(\xi, \rho) \tag{2}$$

where $\tilde{I}^4_4$ (respectively $\tilde{I}^4_5$) is the generalized scalar one loop four point function (resp. the generalized scalar one loop five point function). Here, a generalized scalar one loop N point function differs from a scalar one loop N point function by the integration volume over the Feynman parameters: it is not always the usual hyper-simplex, it can be an hyper-cube or an hyper-prism depending on the topology of the scalar two loop diagram.

As a one loop five point function can be expressed in term of one loop four point functions irrespective of the Feynman parameter phase space volume : $\tilde{I}^4_5 = \sum_i c_i \tilde{I}^4_4$, so one expects that the scalar two loop three and four point functions can be calculated numerically as a sum of two dimensional integrals of generalized one loop scalar four point functions. Besides the integration volume issue, there is another problem to perform this program. Indeed, we cannot use the usual four point function library (FF [1], LOOPTOOLS [2], D0C [3], OLO [4], ...) because they return a result only if the kinematics defining the four point function is in the physical domain. In our case, the kinematics will depend on the two extra integration variables ξ and ρ, and so there is no guarantee that it will belong to such a domain.

So the first step to start this program is to have a scalar four point function formula which is valid outside the physical domain. This is the object of this work. Furthermore, there are very few independent methods for the computation of scalar one loop four point functions in the massive case proposed [5], [6], despite the fact that there are many numerical implementations of these formulae. But for our purpose, the analytic continuation of ref. [5] with such a kinematics remains puzzling and we have had to develop a completely novel method to compute one loop scalar Feynman integrals in a more open way. We will focus here on the case of the four point function but the method can be applied to two or three point functions too.

2. Definitions

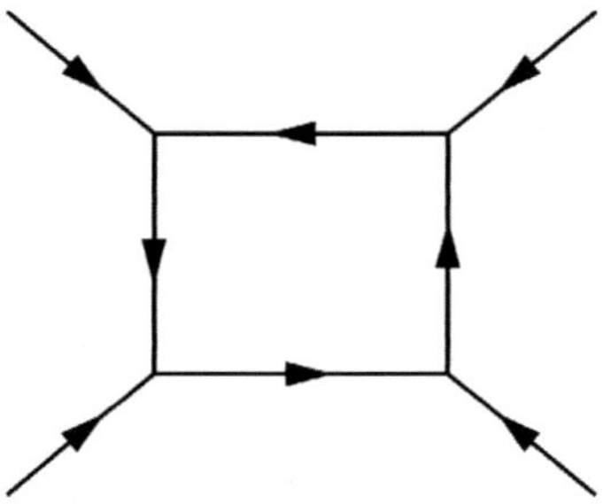

Figure 1. The box picturing the one-loop four point function.

A scalar one loop four point function, depicted in fig. 1, is given, after the integration over the loop momentum, by:

$$I_4^4 = \int_0^1 \prod_{i=1}^{4} dx_i \, \delta(1 - \sum_{j=1}^{4} x_j) \left(-\frac{1}{2} \tilde{X}^{\,T} \mathcal{S} \tilde{X} - i\,\lambda \right)^{-2} \tag{3}$$

where the $\mathcal{S}$ matrix contains all the information on the kinematics and can be expressed in term of the scalar product of the propagator 4-momenta and the internal masses:

$$\mathcal{S}_{i\,j} = (q_i - q_j)^2 - m_i^2 - m_j^2 \tag{4}$$

The raw vector $\tilde{X}$ contains the Feynman parameters as component:

$$\tilde{X} = \begin{bmatrix} x_1 \\ \vdots \\ x_4 \end{bmatrix}$$

and the superscript T stands for the matrix transpose.

IOP Conf. Series: Journal of Physics: Conf. Series **920** (2017) 012001 doi:10.1088/1742-6596/920/1/012001

Expressing x_4 in term of the other Feynman parameters to get rid of the δ constraint, we can write the scalar one loop four point function as:

$$I_4^4 = \int_0^1 dx_1 \int_0^{1-x_1} dx_2 \int_0^{1-x_1-x_2} dx_3 \frac{1}{(D(X) - i\,\lambda)^2} \tag{5}$$

with

$$D(X) = X^{\,T}.A.X + B^T.X + C, \quad X = \begin{bmatrix} x_1 \\ x_2 \\ x_3 \end{bmatrix}$$

The components of the 3×3 matrix A, the three dimensional vector B as the scalar C can be expressed in term of the components of the $\mathcal{S}$ matrix

$$A_{i\,j} = -\frac{1}{2}\,(\mathcal{S}_{i\,j} - \mathcal{S}_{4\,j} - \mathcal{S}_{i\,4} + \mathcal{S}_{4\,4}) \quad i, j \neq 4 \tag{6}$$

$$B_j = \mathcal{S}_{4\,4} - \mathcal{S}_{4\,j} \quad j \neq 4 \tag{7}$$

$$C = -\frac{1}{2}\mathcal{S}_{4\,4} \tag{8}$$

It is important to remark that the matrix A is proportional to a Gram matrix defined in term of the 4-vectors r_i ($q_i = k + r_i$ where k is the momentum running into the loop). More precisely, $A = 1/2\,G^{(4)}$. The superscript 4 indicates that we single out the row and the column 4 of the $\mathcal{S}$ matrix. If another number for a row and a column is singled out, A will be proportional to another Gram matrix which has the same determinant as $G^{(4)}$. As the final result will depend on these Gram matrices only through their determinant, we will drop thereafter the superscript indicating which row and column have been singled out.

3. Description of the method

We will express the three dimensional integral over the simplex as a sum of three dimensional integrals on one eighth of $\mathbb{R}^3$. In order to achieve that, we will make an extensive use of the following identity[1]:

$$\frac{1}{D^{\alpha+1}} = \frac{1}{\alpha\,\Delta_n}\left[2\,\frac{n-2\,\alpha}{D^\alpha} - \nabla^T.\left(\frac{2\,X + A^{-1}.B}{D^\alpha}\right)\right] \tag{9}$$

where n is the number of independent integration variables (here 3) and:

$$\Delta_n = B^T.A^{-1}.B - 4\,C$$

The eq. (9) is interesting to use in the case where $\alpha = n/2$ because in such a case only the boundary term remains. But, in general, $\alpha \neq n/2$, for example for the four point function $\alpha = 1$ and $n = 3$. The idea is to introduce a new identity which enables to adjust the power of the denominator of eq. (5). For that, we introduce the following relation valid for $\mu > 1/\nu > 0$ and any complex number D:

$$\int_0^\infty \frac{d\xi}{(D + \xi^\nu)^\mu} = \frac{1}{\nu}\,B\left(\frac{1}{\nu}, \mu - \frac{1}{\nu}\right)\,\frac{1}{D^{\mu-1/\nu}} \tag{10}$$

where $B(x, y)$ is the Euler beta function defined by

$$B(x, y) = \frac{\Gamma(x)\,\Gamma(y)}{\Gamma(x + y)}$$

[1] Integrating this equality over the simplex gives the usual relations between scalar one loop four point function in different space-time dimensions [7],[8]

A first step consists of an adjustment of the power of the denominator in the l.h.s of eq. (9) in such way that only the boundary term of eq. (9) remains. In the case of the four point function, cf. eq. (5) $n = N - 1$ is equal to 3. Imposing $n - 2\alpha$ to vanish thus forces α to be $3/2$, hence a power $\alpha + 1 = 5/2$ in the l.h.s. of eq. (9). However in eq. (5), D is raised to the power 2, not $5/2$. In order to shift the power of the denominator which arises naturally in eq. (5), by $1/2$, we use identity (10) i.e. the power of D appearing in the denominator of the integrand of eq. (5). The representation of $1/D^2$ thus obtained is substituted into eq. (5). This provides a representation of I_4^4 in whose integrand D is replaced by $(D+\xi^\nu)$. Identity (9) is then applied to *this* new integrand considered as a function of the three integration variables x_1, x_2, x_3 whereas ξ is seen as a fixed parameter. In order that the first term of the thereby modified (9) be made vanishing, μ is chosen to be $3/2$ thus ν shall be chosen equal to 2. Eq. (5) thus reads:

$$\begin{aligned} I_4^4 &= \int_0^\infty d\xi \, \frac{2}{B(2,1/2)} \int_0^1 dx_1 \int_0^{1-x_1} dx_2 \\ &\times \int_0^{1-x_1-x_2} dx_3 \, \frac{1}{(D(x_1,x_2,x_3) + \xi^2 - i\,\lambda)^{5/2}} \end{aligned} \tag{11}$$

and a direct application of eq. (9), gives:

$$\begin{aligned} I_4^4 &= \frac{1}{3\,B(2,1/2)} \int_0^\infty d\xi \, \frac{1}{\xi^2 - \Delta_3/4 - i\,\lambda} \int_0^1 dx_1 \int_0^{1-x_1} dx_2 \\ &\times \int_0^{1-x_1-x_2} dx_3 \, \nabla^T . \left(\frac{2\,X + A^{-1}.B}{(D(x_1,x_2,x_3) + \xi^2 - i\,\lambda)^{3/2}} \right) \end{aligned} \tag{12}$$

with $\Delta_3 = B^T.A^{-1}.B - 4\,C$.

Note that we have to be careful because we are working with complex numbers and it is not obvious that the integration over ξ in eq. (12) is well defined. More precisely, we can ask ourselves what happens if $\mathrm{Im}(\Delta_3) \to 0$ and so the poles in ξ lie on the real axis. To answer such a question, we have to study what is the residue of the integrand of eq. (12) when $\xi = \pm\sqrt{\Delta_3/4}$[2]. The residue of the poles in ξ, named R, is given by :

$$R = \int_0^1 dx_1 \int_0^{1-x_1} dx_2 \int_0^{1-x_1-x_2} dx_3 \, \nabla^T . \left(\frac{2\,X + A^{-1}.B}{(D(X) + \Delta_3/4)^{3/2}} \right) \tag{13}$$

If $\mathrm{Im}(\Delta_3) = 0$, the imaginary part of the denominator is driven by $\mathrm{Im}(D(X))$ which is of constant sign when the Feynman parameters travel across the simplex, that means that the function $(D(X) + \Delta_3/4)^{3/2}$ never crosses its cut in this domain. So $(2\,X + A^{-1}.B)/(D(X) + \Delta_3/4)^{3/2}$ is a gentle function when X travels across the simplex. It is easy to show that, for such a function,

$$\nabla^T \left(\frac{2\,X + A^{-1}.B}{(D(X) + \Delta_3/4)^{3/2}} \right) = 0 \tag{14}$$

So for $\mathrm{Im}(\Delta_3) = 0$, $R = 0$ and the values $\xi = \pm\sqrt{\Delta_3/4}$ are not poles for the integrand of eq. (12).

By computing explicitly the gradient in (12) and rearranging different terms, we end up with:

$$\begin{aligned} I_4^4 &= -\frac{2}{3} \frac{1}{B(2,1/2)} \int_0^\infty d\xi \, \frac{1}{\xi^2 - \Delta_3/4 - i\,\lambda} \sum_{i=1}^{4} \frac{\bar{b}_i}{\det(G)} \\ &\times \int_0^1 dx_1 \int_0^{1-x_1} dx_2 \, \frac{1}{(\overline{D}_i(x_1,x_2) + \xi^2 - i\,\lambda)^{3/2}} \end{aligned} \tag{15}$$

[2] We skip the small imaginary part because this discussion is relevant only in the case of complex masses.

where we have introduced the following quantities :

$$\begin{aligned}\Delta_3 &= -2\,\frac{\det(\mathcal{S})}{\det(G)}\\ \bar{b}_i &= \det(\mathcal{S})\sum_{j\in S}\mathcal{S}^{-1}_{i\,j}\end{aligned}$$

The new quadratic forms $\overline{D}_i(x_1,x_2)$ are related to the original one by:

$$\begin{array}{ll}\overline{D}_1(x_1,x_2)=D(0,x_1,x_2) & \overline{D}_2(x_1,x_2)=D(x_2,0,x_1)\\ \overline{D}_3(x_1,x_2)=D(x_1,x_2,0) & \overline{D}_4(x_1,x_2)=D(1-x_1-x_2,x_1,x_2)\end{array}$$

A second step consists of the application of the same method to the integration over the two left Feynman parameters x_1 and x_2 in (15) : the new quadratic forms $\overline{D}_i(x_1,x_2)$ can be written as :

$$\overline{D}_i(x_1,x_2)=\bar{X}^{\,T}.E_i.\bar{X}+F_i^T.\bar{X}+G_i$$

where E_i is a 2×2 matrix, F_i a two dimensional column vector and G_i a scalar. The two dimensional vector $\bar{X}$ is defined by :

$$\bar{X}=\begin{bmatrix}x_1\\ x_2\end{bmatrix}$$

In order that only the boundary term remains when applying eq. (9), we have to shift the power of the denominator and thus to introduce a new integral representation for it. After the second step, we end up with the integration on the left over Feynman parameter of a new quadratic form.

The third step, using the same method, will trade this remaining Feynman parameter integration with an integration over σ on half $\mathbb{R}$, σ being the variable introduced to shift the power of the denominator in the last step.

We skip all the details and, after step 2 and 3, the final result is given by:

$$\begin{aligned}I_4^4 &= \frac{4}{3}\,\frac{1}{B(2,1/2)}\,\frac{1}{B(3/2,1/2)}\,\frac{1}{B(1,1/2)}\sum_{i\in S}\;\sum_{j\in S\setminus\{i\}}\;\sum_{k\in S\setminus\{i,j\}}\\ &\times\frac{\bar{b}_i}{\det(G)}\,\frac{\bar{b}_j^{\{i\}}}{\det(G^{\{i\}})}\,\frac{\bar{b}_k^{\{ij\}}}{\det(G^{\{ij\}})}\int_0^\infty d\xi\int_0^\infty d\rho\int_0^\infty d\sigma\\ &\times\frac{1}{\xi^2-\Delta_3/4-i\,\lambda}\,\frac{1}{\xi^2+\rho^2-\Delta_2^i/4-i\,\lambda}\\ &\times\frac{1}{\xi^2+\rho^2+\sigma^2-\Delta_1^{ij}/4-i\,\lambda}\,\frac{1}{(\tilde{D}_{ijk}+\xi^2+\rho^2+\sigma^2-i\,\lambda)^{1/2}}\end{aligned} \quad (16)$$

Some explanations are needed in order to understand the notations of eq, (16). We introduce a set $S=\{1,2,3,4\}$ which contains the labels of the propagator of the four point functions. At the first step, we work with a 4×4 $\mathcal{S}$ matrix from which we extracted a 3×3 Gram matrix G. At the second step, the quadratic forms $\overline{D}_i(x_1,x_2)$ are the quadratic forms obtained by pinching one of the propagator of the initial diagram to which corresponds a 3×3 $\mathcal{S}$ matrix named $\mathcal{S}^{\{i\}}$ and which is obtained from the initial $\mathcal{S}$ matrix by removing the row and the column i. From this reduced $\mathcal{S}$ matrix, we can extract a reduce Gram matrix named $G^{\{i\}}$. At the third step, we work with one dimensional quadratic forms which are the one obtained by pinching two propagators of the initial diagram, this leads to a 2×2 $\mathcal{S}$ matrix named $\mathcal{S}^{\{i,j\}}$ obtained

IOP Conf. Series: Journal of Physics: Conf. Series **920** (2017) 012001 doi:10.1088/1742-6596/920/1/012001

by removing the rows and the columns i and j. Again, from that matrix we can extract a one dimensional Gram matrix named $G^{\{i,j\}}$. At any step, we never performed a relabelling of the subscripts describing the left rows and columns of the $\mathcal{S}$ matrix, this is for this reason that the subscripts j and k run over a subset of the set S.

At the end of the steps two and three, the following quantities have appeared which are defined as:

$$\tilde{D}_{ijk} = m_l^2 \quad \text{with} \quad l = S \setminus \{i\, j\, k\}$$

and

$$\Delta_1^{ij} = -\frac{2\,\det\left(\mathcal{S}^{\{i,j\}}\right)}{\det\left(G^{\{i,j\}}\right)}$$

$$\Delta_2^{i} = \frac{2\,\det\left(\mathcal{S}^{\{i\}}\right)}{\det\left(G^{\{i\}}\right)}$$

4. Integration

It remains to perform the three integrations over ξ, ρ and σ.

$$\begin{aligned} L_{ijk} &= -\frac{\kappa}{4^3} \int_0^\infty d\xi \int_0^\infty d\rho \int_0^\infty d\sigma \\ &\times \frac{1}{\xi^2 - \frac{\Delta_3}{4} - i\,\lambda} \, \frac{1}{\xi^2 + \rho^2 - \frac{\Delta_2^i}{4} - i\,\lambda} \\ &\times \frac{1}{\xi^2 + \rho^2 + \sigma^2 - \frac{\Delta_1^{ij}}{4} - i\,\lambda} \, \frac{1}{(\tilde{D}_{ijk} + \xi^2 + \rho^2 + \sigma^2 - i\,\lambda)^{1/2}} \end{aligned}$$

with

$$\kappa = -\frac{256}{3} \, \frac{1}{B(2,1/2)} \, \frac{1}{B(3/2,1/2)} \, \frac{1}{B(1,1/2)}$$

To do that, we will prove the following intermediate result. Let us consider the following integral:

$$J = \int_0^\infty d\xi \, \frac{1}{\xi^2 + A} \, \frac{1}{(\xi^2 + B)^{1/2}} \tag{17}$$

where A and B are complex numbers. We have to distinguish two cases depending on S_A and S_B: the sign of the imaginary part of A (respectively B). Using Feynman parameter method, we can show that:

$$J = \int_0^1 dz \, \frac{1}{z^2\, B + (1 - z^2)\, A} \tag{18}$$

if $S_A = S_B$ and:

$$J = i\, S_B \int_0^\infty \frac{dz}{B\, z^2 - (1 + z^2)\, A} - \int_1^\infty \frac{dz}{B\, z^2 + (1 - z^2)\, A} \tag{19}$$

otherwise.

Let us sketch the way we proceed.

- For the σ integration: this integration is of the type (17), so we use eq. (18) or (19) to trade the sigma integration with a z integration.

- For the ρ integration: this integration can be done using partial fraction decomposition and eq. (10).

- For the ξ integration: again, this integration is of the type (17) and can be trade, using eq. (18) or (19), with a v integration.

To go further, we have to distinguish the different cases, there are 8 cases to be considered. We present here, as example, only two cases.

4.1. Case $\mathrm{Im}(\Delta_3) > 0$, $\mathrm{Im}(\Delta_2^i) > 0$, $\mathrm{Im}(\Delta_1^{ij} > 0$, $\mathrm{Im}(\tilde{D}_{ijk}) < 0$

This corresponds to the real mass case. After the integration over ξ, ρ and σ, we end up with:

$$L_{ijk} = \frac{1}{2}\int_0^1 \frac{dv}{v^2\,P_{ijk} + R_{ij}} \left[\int_0^1 \frac{dz}{z^2\,(T + Q_i) + (1 - z^2)\,T} - \int_0^1 \frac{dz}{z^2\,(v^2\,P_{ijk} + R_{ij} + Q_i + T) + (1 - z^2)\,T}\right] \tag{20}$$

where we have introduced the following quantities:

$$P_{ijk} = \tilde{D}_{ijk} + \frac{\Delta_1^{ij}}{4} \tag{21}$$

$$Q_i = \frac{\Delta_3}{4} - \frac{\Delta_2^i}{4} \tag{22}$$

$$R_{ij} = \frac{\Delta_2^i}{4} - \frac{\Delta_1^{ij}}{4} \tag{23}$$

$$T = -\frac{\Delta_3}{4} \tag{24}$$

Then, we made the following change of variables $x = \sqrt{z}$ and $y = \sqrt{v\,z}$ and exchange the order of the y and x integration. After these steps, the x integration is easy to perform and setting $u^2 = y$, we get the following result:

$$L_{ijk} = -\frac{1}{4}\int_0^1 du\,\frac{1}{u^2\,P_{ijk}\,Q_i - R_{ij}\,T}\left[\ln\left(\frac{u^2\,P_{ijk} + R_{ij} + Q_i + T}{u^2\,(P_{ijk} + R_{ij} + Q_i) + T}\right) - \ln\left(\frac{Q_i + T}{u^2\,Q_i + T}\right)\right] \tag{25}$$

As an illustration, let us consider another case.

4.2. Case $\mathrm{Im}(\Delta_3) < 0$, $\mathrm{Im}(\Delta_2^i) > 0$, $\mathrm{Im}(\Delta_1^{ij}) > 0$, $\mathrm{Im}(\tilde{D}_{ijk}) < 0$

For this case, we will give no intermediate steps and the final results is given by:

$$\begin{aligned} L_{ijk} = \frac{1}{4}\Bigg\{ & i\int_0^\infty du\,\frac{1}{u^2\,P_{ijk}\,Q_i + R_{ij}\,T}\left[\ln\left(\frac{R_{ij} + Q_i}{u^2\,(P_{ijk} + R_{ij} + Q_i) - T}\right) - \ln\left(\frac{Q_i}{Q_i\,u^2 - T}\right)\right] \\ & + \int_1^\infty du\,\frac{1}{u^2\,P_{ijk}\,Q_i - R_{ij}\,T}\left[\ln\left(\frac{R_{ij} + Q_i}{u^2\,(P_{ijk} + R_{ij} + Q_i) + T}\right) - \ln\left(\frac{Q_i}{Q_i\,u^2 + T}\right)\right] \\ & + \int_0^1 du\,\frac{1}{u^2\,P_{ijk}\,Q_i - R_{ij}\,T}\left[\ln\left(\frac{R_{ij} + Q_i}{u^2\,P_{ijk} + R_{ij} + Q_i + T}\right) - \ln\left(\frac{Q_i}{Q_i + T}\right)\right]\Bigg\} \end{aligned} \tag{26}$$

The leftover u integration is standard to perform and can be expressed in term of Li_2 and logarithm functions.

Nice features of the method are that for each cases (for instance eq. (25) or eq. (26)), the residue at the poles $u^2 = R_{ij}\,T/(P_{ijk}\,Q_i)$ or $u^2 = -R_{ij}\,T/(P_{ijk}\,Q_i)$ is zero by construction and also that for each integral, the logarithms never cross their cuts when u travels along the path on the real axis defined by the bounds of the integration.

5. Conclusion

We presented a novel method to compute the scalar one loop four point functions, this method enables to compute four point functions with kinematics outside the physical domain in a very open way. In addition, the results are expressed in term of the determinant of the $\mathcal{S}$ matrix and reduced $\mathcal{S}$ matrices and also in term of the determinant of the Gram matrix and reduced Gram matrices. The drawback of this method is that it leads at the end to a formula which has more Li_2 than the other formula [5]. This method extends rather easily to the case where there are collinear/soft divergences. The results of all these formula have been coded into a FORTRAN 95 program and checked against the existing results [4], [9].

References

[1] van Oldenborgh G J 1991 *Comput. Phys. Commun.* **66** 1–15
[2] Hahn T and Perez-Victoria M 1999 *Comput. Phys. Commun.* **118** 153–165 (*Preprint* hep-ph/9807565)
[3] Nhung D T and Ninh L D 2009 *Comput. Phys. Commun.* **180** 2258–2267 (*Preprint* 0902.0325)
[4] van Hameren A 2011 *Comput. Phys. Commun.* **182** 2427–2438 (*Preprint* 1007.4716)
[5] 't Hooft G and Veltman M J G 1979 *Nucl. Phys.* **B153** 365–401
[6] Denner A, Nierste U and Scharf R 1991 *Nucl. Phys.* **B367** 637–656
[7] Tarasov O V 1996 *Phys. Rev.* **D54** 6479–6490 (*Preprint* hep-th/9606018)
[8] Binoth T, Guillet J P and Heinrich G 2000 *Nucl. Phys.* **B572** 361–386 (*Preprint* hep-ph/9911342)
[9] Cullen G, Guillet J P, Heinrich G, Kleinschmidt T, Pilon E, Reiter T and Rodgers M 2011 *Comput. Phys. Commun.* **182** 2276–2284 (*Preprint* 1101.5595)

IOP Conf. Series: Journal of Physics: Conf. Series **920** (2017) 012002 doi:10.1088/1742-6596/920/1/012002

Efficient reduction of four-loop massless propagators

T Ueda[1], B Ruijl[1,2] and J A M Vermaseren[1]

[1] Nikhef Theory Group, Science Park 105, 1098 XG Amsterdam, The Netherlands
[2] Leiden Centre of Data Science, Leiden University, Niels Bohrweg 1, 2333 CA Leiden, The Netherlands

E-mail: `tueda@nikhef.nl`, `benrl@nikhef.nl`, `t68@nikhef.nl`

Abstract. In the light of precision measurements at the LHC experiments, theory predictions including higher-order corrections have become more important. We have developed a program `Forcer` dedicated to analytical evaluation of four-loop massless propagator-type Feynman integrals via parametric integration-by-parts reduction. Some recent physics results obtained by `Forcer` are discussed.

1. Introduction

In the last decade, much effort has been made on accurate theory predictions to meet requirements from precision experiments. Especially, precise QCD predictions are crucial for experiments at hadron colliders due to the strong interactions, hence theory calculations for higher-order radiative corrections in perturbative theory are mandatory to analyze a large amount of data from detectors.

A notoriously well-known example that slowly converges and hence requires higher-order calculations is the Higgs production cross section in the gluon-fusion channel $gg \to H$, which is the dominant channel in hadron colliders. In this channel, the next-leading order corrections in the strong coupling give a large contribution comparable to that from the leading order, thus in practice one has to use the next-to-next-to leading order (NNLO) prediction for any phenomenological studies. Recently, a more precise result for the total cross section including the next-to-next-to-next-to leading order (N^{3}LO) corrections in the heavy top-quark mass limit became available [1], which reduced the total uncertainty down to less than 10% for the LHC at 13 TeV.

One of the remaining sources of theory uncertainty for the total Higgs production cross section originates from the fact that parton distribution functions are extracted from experimental data via NNLO analysis while the partonic cross section is obtained up to the N^{3}LO. In order to improve this situation, one must know four-loop splitting functions, however, a direct computation of them sounds too challenging at present. A more promising approach in the near-to-mid term is to compute low-integer Mellin N-th moments of the splitting functions and estimate them via, e.g., Padé approximant. Indeed, such estimations were performed at the three-loop level [2, 3, 4] before the full N-dependence was computed [5, 6]. At the four-loop level, until recently, these computations had been done only up to $N = 4$ [7, 8, 9, 10].

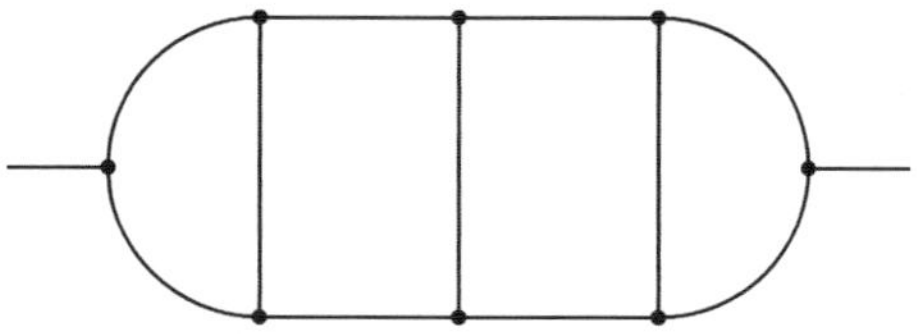

(a) ladder

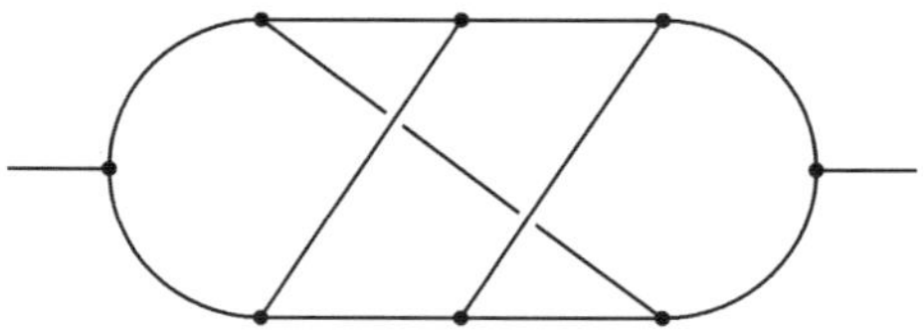

(b) non-planar (no1)

Figure 1. Two topologies out of 11 top-level topologies at the four-loop level.

Once N is fixed to an integer, such an N-th moment calculation reduces to evaluation of massless propagator-type Feynman integrals. Master integrals in this class of integrals are known at the four-loop level [11, 12]. The main obstacle for computing moments for higher N is, however, the reduction to the master integrals via integration-by-parts (IBP) identities [13, 14] within dimensional regularization [15, 16]. Although there are several generic IBP reduction methods available on the market — for example, Laporta's algorithm [17] (see [18, 19, 20, 21, 22, 23] for public implementations), Baikov's method [24, 25] and Lee's heuristic approach [26, 27] — it is still a tough problem to perform reductions needed for higher moments in a reasonable amount of computational time.

On the other hand, up to the three-loop level, it is known that many massless propagator-type integrals can be reduced into simpler ones by using the so-called triangle rule [13, 14, 28] constructed from the IBP identities and by performing one-loop massless integrations with G-functions [29, 14]. There are only two topologies that require manually solved IBP reduction rules [14]. In this way, the IBP reduction need not be performed when one arrives at integrals that can be expressed in terms of the gamma function. This approach was implemented in the `Mincer` program [30] written in `SCHOONSCHIP` [31] and later reprogrammed [32] in `FORM` [33, 34], which is highly efficient and has been used for many calculations including very high moments (e.g., [35, 36]).

The successful experience at the three-loop level suggests that if one could extend the `Mincer` program to the four-loop level then it might be more efficient than general IBP solvers for massless propagator-type integrals. This is a natural expectation because usually an algorithm with domain-specific knowledge has a better performance than a generic algorithm. This work aims to develop a new `FORM` package `Forcer` [37, 38, 39], which is, in a word, an extension of `Mincer` to the four-loop level, and push the limit of our capability for this class of integrals. Due to a high complexity of the program at the four-loop level, code generation must be as much automatized as possible. `Forcer` allows us to compute more high N moments of four-loop splitting functions and can be useful for other calculations (see [10] for a recent review on physics related to massless propagators).

2. Reduction of each topology by its substructure

Families of scalar Feynman integrals can be classified by their denominator structure or topologies of corresponding Feynman diagrams. In our case, all (squared) propagators appearing in the denominator are massless. Figure 1 shows example topologies at the four-loop level. Generic (top-level) topologies, consisting of three-point vertices, at this level have 11 internal lines. There are also 3 irreducible numerators, which cannot be expressed in terms of propagators, and hence they are characterized by 14 parameters (called as indices).

As seen in the `Mincer` approach, some substructures found in topologies lead to reduction rules of integrals belonging to a topology into ones in simpler topologies.

IOP Conf. Series: Journal of Physics: Conf. Series **920** (2017) 012002 doi:10.1088/1742-6596/920/1/012002

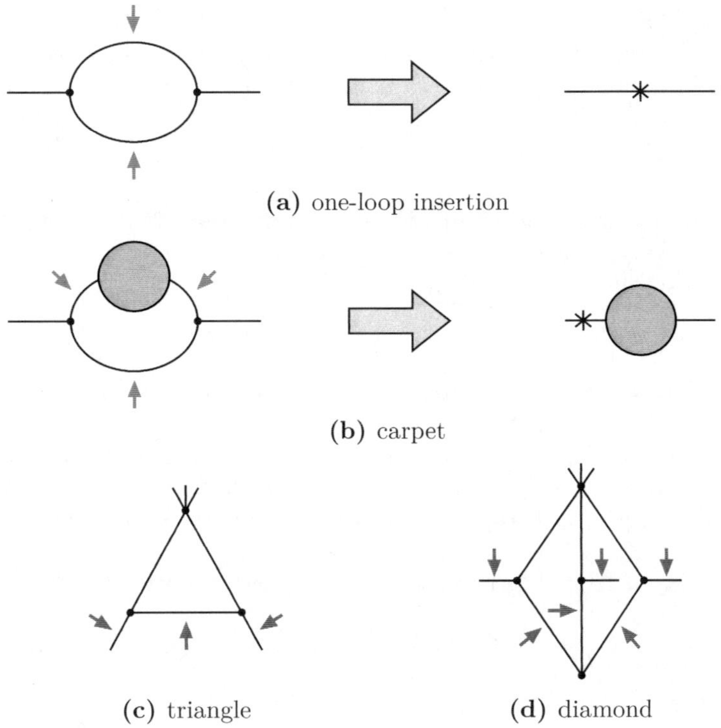

Figure 2. Substructures of topologies that give reductions of integrals into ones in simpler topologies: (a) one-loop insertion, (b) carpet, (c) triangle and (d) two-loop diamond. A line with a non-integer power is indicated with "$*$". Performing a one-loop insertion integral or carpet integral removes the lines indicated by the arrows. The triangle rule and diamond rule remove one of the lines indicated by the arrows.

2.1. One-loop insertion integral

When a topology contains a massless one-loop insertion graph (figure 2a) as its substructure, one can integrate out the corresponding loop momentum. The integration gives G-functions, which can be expressed in terms of the gamma function and are exact in $D = 4 - 2\epsilon$ space-time dimensions. Graphically the two lines in the loop become a single line but it gets a non-integer power $1/(Q^2)^\epsilon$ of the resultant momentum Q (indicated by "$*$" in figure 2a). The powers of the two propagators can be arbitrary numbers.

2.2. Carpet integral

There is another type of massless one-loop integrals that can be easily performed. When a subgraph is inserted to a one-loop graph (figure 2b), the outer loop can be integrated out first. The powers of the two outer propagators can be arbitrary numbers.

2.3. Triangle rule

When a topology contains a one-loop triangle subgraph depicted in figure 2c, there is a combination of IBP identities that decreases a sum of powers of the three propagators indicated by the arrows (triangle rule). If the powers of the three propagators are positive integers, then recursive use of the triangle rule reduces an integral into a sum of integrals in simpler topologies with one of the three propagators removed. The number of external lines attached to the top vertex has no restriction. For example, the ladder topology in figure 1a has two triangles where the triangle rule is applicable. Note that non-integer powers of propagators can be introduced by one-loop insertion integrals or carpet integrals and they may disallow a use of the triangle rule.

2.4. Diamond rule

The triangle rule has an extension for multi-loop diamond-shaped subgraphs [40]. Figure 2d shows the simplest diamond structure, which appears at the four-loop level for the first time and removes one of the six lines indicated by the arrows.

2.5. Special rule

In the non-planar topology in figure 1b (referred as "no1" in `Forcer`), one cannot find any of the above substructures. For a topology where none of the useful substructures is available, one has to prepare "manually" solved IBP reduction rules in such a way that integrals are reduced to ones in simpler topologies or at least to the master integral in the topology. We used considerably computer-assisted derivations of IBP reduction rules for each topology, although it has not been fully automatized.

We start from a set S_0 of IBP identities, which are constructed in the usual way,

$$S_0 = \{I_1, \ldots, I_{R_0}\}. \tag{1}$$

Each IBP identity, given as $I_i = 0$, is a linear relation of integrals with various indices. Their coefficients contain the regulator ϵ of dimensional regularization and the indices n_j parametrically. The number of the relations is $R_0 = L(L+E)$ for an L-loop topology with E-independent external momenta. We make a set S_1 of relations obtained from S_0 with an index shifted by one in all possible ways. The set S_1 has $R_1 = NR_0$ elements for a topology with N-indices. Many useful reduction rules can be found as a linear combination of IBPs in the combined set of $S_0 \cup S_1$:

$$\sum_i c_i I_i = 0, \qquad I_i \in S_0 \cup S_1. \tag{2}$$

Basically this is achieved by elimination of complicated integrals in the system. We have constructed an reduction scheme for each topology by combining resulted identities.

3. Construction of the reduction flow for all topologies

In the `Mincer` program, all topologies appearing in the problem were classified by their substructure and all the cases, requiring reductions discussed in section 2, were programmed by hand. This means that the program also contains code for transition from a topology to simpler topologies, for example, rewriting a missing momentum that does not exist in a simpler topology and applying topology symmetries that equate several topologies and so on. At the four-loop level, however, it becomes highly complicated and error-prone. Therefore one should completely automatize the code generation for handling each topology and transition among topologies as well as the classification.

We represent a topology as an undirected graph in graph theory. This makes it easy to detect substructure in a graph; the classification can be done just by pattern matchings of connections of vertices and edges. The reason why we use undirected graphs is that all propagators are squared p_i^2 and after integration the result is a function of the square of the external momentum Q^2, so their directions do not matter. Undirected graphs exhibit symmetries for such momentum directions. One tricky point is the fact that we use dot products ($p_i \cdot p_j$ or $p_i \cdot Q$) as irreducible numerators, hence an extra minus sign may have to be compensated when a momentum is flipped. Non-integer powers of propagators are treated by assigning "colors" to edges, which enables us to distinguish different numbers of non-integer powers m_i of $1/(p_i^2)^{n_i+m_i\epsilon}$. Massive propagators may be handled by additional colors of edges although all propagators are massless in our problem.

The reduction flow of all topologies is constructed as follows. Starting from the top-level topologies, we consider all possible ways to remove a line for each topology. Topologies with massless tadpoles are immediately discarded. Some of the generated topologies are indeed identical under graph isomorphism, which is detected by graph algorithms. We keep track of all mappings of momenta in topology transitions, which is necessary for rewriting propagators and irreducible numerators. For each topology, the next action is determined by its substructure. Irreducible numerators are chosen in such a way that they do not interfere with the action. Symmetries in a topology are detected as graph automorphisms, which helps to reduce the number of terms in the calculation. We repeat this procedure until all topologies are reduced into the Born graph, which creates the whole reduction flow of all possible topologies in the problem.

We have implemented this method by using `Python` with a graph library `igraph` [41]. The program generates `FORM` code in the end. From 11 top-level topologies at the four-loop level, 437 non-trivial topologies appear in the reduction flow. The numbers of topologies with one-loop insertions, carpet integrals, triangles and diamonds are 335, 24, 53 and 4, respectively. The remaining 21 topologies require special rules.

4. Discussion

Together with the special rules and a set of auxiliary library routines, the `FORM` code generated as in section 3 leads to the `Forcer` package. To give a feeling about the performance of `Forcer`, we performed a benchmark reduction of $\mathbf{no1}(2,2,2,2,2,2,2,2,2,2,2,-1,-1,-1)$, which is obtained by increasing indices for all propagators and decreasing them for all irreducible numerators from the master integral $\mathbf{no1}(1,1,1,1,1,1,1,1,1,1,1,0,0,0)$ in the "no1" topology (figure 1b). It took about 4 hours on a desktop PC with 4 cores. This can be improved if one uses a truncated expansion of ϵ in intermediate steps rather than rational arithmetic.

In practice, one needs to compute Feynman diagrams, which must be also automatically generated in some way. Our setup uses `QGRAF` [42] for generating diagrams. The output is then manipulated in `FORM`, as discussed in Ref. [43], and passed to `Forcer`.

The correctness of the program was checked by recomputing several known results found in the literature. Especially, an computation of the four-loop QCD beta function [44, 45] via renormalization constants of self-energies and a vertex with keeping the full dependence of the gauge parameter provides a strong check (the final result of the beta function is of course gauge independent in the minimum subtraction scheme). Others include lower moments of the four-loop splitting functions in the flavor non-singlet sector [7, 8, 9, 10].

`Forcer` was successfully used for several calculations. In Refs. [46, 47], Mellin moments of the four-loop splitting functions and coefficient functions were studied. They include the moments of the splitting functions up to $N = 6$ and $N = 4$ for the non-singlet and singlet cases, respectively, and the moments for structure functions $F_{2,ns}$ and $F_{L,ns}$ up to $N = 6$. Interestingly, the full N-dependence of n_f^2 contributions to non-singlet splitting functions and n_f^3 contributions to singlet splitting functions at this order can be completely determined by computing a fairly large number of Mellin moments and using the LLL algorithm [48]. In Ref. [49], by combining `Forcer` with the R^*-operation [50, 51, 52, 53], the five-loop beta function of Yang-Mills theory coupled to fermions for a general gauge group was obtained, which extended and hence confirmed results obtained in Refs. [54, 55]. More physics results by `Forcer` will be reported elsewhere [56].

The construction of the reduction flow described in section 3 is applicable also at the five-loop level. There are 6570 topologies obtained from 64 top-level topologies. However, 284 topologies have no useful substructure and they require special reduction rules. At present, this seems rather difficult and probably it requires complete automatization for solving IBP relations. Furthermore, derivation of parametric reduction rules can be helpful for other processes. Even partial reductions could be used for preprocessing input integrals in other generic IBP solvers, e.g., Laporta's algorithm, and might give a performance gain. We leave these problems for future work.

Acknowledgments

We would like to thank Joshua Davies, Franz Herzog and Andreas Vogt for discussions and collaboration for physics applications. This work is supported by the ERC Advanced Grant no. 320651, "HEPGAME". Figures were drawn with `Axodraw2` [57].

References

[1] Anastasiou C, Duhr C, Dulat F, Furlan E, Gehrmann T, Herzog F, Lazopoulos A and Mistlberger B 2016 *JHEP* **05** 058 (*Preprint* 1602.00695)
[2] Larin S A, van Ritbergen T and Vermaseren J A M 1994 *Nucl. Phys.* **B427** 41–52
[3] Larin S A, Nogueira P, van Ritbergen T and Vermaseren J A M 1997 *Nucl. Phys.* **B492** 338–78 (*Preprint* hep-ph/9605317)
[4] Retey A and Vermaseren J A M 2001 *Nucl. Phys.* **B604** 281–311 (*Preprint* hep-ph/0007294)
[5] Moch S, Vermaseren J A M and Vogt A 2004 *Nucl. Phys.* **B688** 101–34 (*Preprint* hep-ph/0403192)
[6] Vogt A, Moch S and Vermaseren J A M 2004 *Nucl. Phys.* **B691** 129–81 (*Preprint* hep-ph/0404111)
[7] Baikov P A and Chetyrkin K G 2006 *Nucl. Phys. Proc. Suppl.* **160** 76–9
[8] Velizhanin V N 2012 *Nucl. Phys.* **B860** 288–94 (*Preprint* 1112.3954)
[9] Velizhanin V N 2014 *Preprint* 1411.1331
[10] Baikov P A, Chetyrkin K G and Kühn J H 2015 *Nucl. Part. Phys. Proc.* **261-262** 3–18 (*Preprint* 1501.06739)
[11] Baikov P A and Chetyrkin K G 2010 *Nucl. Phys.* **B837** 186–220 (*Preprint* 1004.1153)
[12] Lee R N, Smirnov A V and Smirnov V A 2012 *Nucl. Phys.* **B856** 95–110 (*Preprint* 1108.0732)
[13] Tkachov F V 1981 *Phys. Lett.* **B100** 65–8
[14] Chetyrkin K G and Tkachov F V 1981 *Nucl. Phys.* **B192** 159–204
[15] Bollini C G and Giambiagi J J 1972 *Nuovo Cim.* **B12** 20–6
[16] 't Hooft G and Veltman M J G 1972 *Nucl. Phys.* **B44** 189–213
[17] Laporta S 2000 *Int. J. Mod. Phys.* **A15** 5087–159 (*Preprint* hep-ph/0102033)
[18] Anastasiou C and Lazopoulos A 2004 *JHEP* **07** 046 (*Preprint* hep-ph/0404258)
[19] Smirnov A V 2008 *JHEP* **10** 107 (*Preprint* 0807.3243)
[20] Smirnov A V and Smirnov V A 2013 *Comput. Phys. Commun.* **184** 2820–7 (*Preprint* 1302.5885)
[21] Smirnov A V 2014 *Comput. Phys. Commun.* **189** 182–91 (*Preprint* 1408.2372)
[22] Studerus C 2010 *Comput. Phys. Commun.* **181** 1293–300 (*Preprint* 0912.2546)
[23] von Manteuffel A and Studerus C 2012 *Preprint* 1201.4330
[24] Baikov P A 1996 *Phys. Lett.* **B385** 404–10 (*Preprint* hep-ph/9603267)
[25] Baikov P A 2006 *Phys. Lett.* **B634** 325–9 (*Preprint* hep-ph/0507053)
[26] Lee R N 2012 *Preprint* 1212.2685
[27] Lee R N 2014 *J. Phys. Conf. Ser.* **523** 012059 (*Preprint* 1310.1145)
[28] Tkachov F V 1983 *Theor. Math. Phys.* **56** 866–70
[29] Chetyrkin K G, Kataev A L and Tkachov F V 1980 *Nucl. Phys.* **B174** 345–77
[30] Gorishny S G, Larin S A, Surguladze L R and Tkachov F V 1989 *Comput. Phys. Commun.* **55** 381–408
[31] Strubbe H 1974 *Comput. Phys. Commun.* **8** 1–30
[32] Larin S A, Tkachov F V and Vermaseren J A M The FORM version of MINCER NIKHEF-H/91-18
[33] Vermaseren J A M 2000 *Preprint* math-ph/0010025
[34] Kuipers J, Ueda T, Vermaseren J A M and Vollinga J 2013 *Comput. Phys. Commun.* **184** 1453–67 (*Preprint* 1203.6543)
[35] Velizhanin V N 2012 *Nucl. Phys.* **B864** 113–40 (*Preprint* 1203.1022)
[36] Moch S, Vermaseren J A M and Vogt A 2014 *Nucl. Phys.* **B889** 351–400 (*Preprint* 1409.5131)
[37] Ueda T, Ruijl B and Vermaseren J A M 2016 *J. Phys. Conf. Ser.* **762** 012060 (*Preprint* 1604.08767)
[38] Ueda T, Ruijl B and Vermaseren J A M 2016 *PoS* **LL2016** 070 (*Preprint* 1607.07318)
[39] Ruijl B, Ueda T and Vermaseren J A M in preparation
[40] Ruijl B, Ueda T and Vermaseren J 2015 *Phys. Lett.* **B746** 347–50 (*Preprint* 1504.08258)

[41] Csardi G and Nepusz T The igraph software package for complex network research *InterJournal* **Complex Systems** (2006) 1695, http://igraph.org
[42] Nogueira P 1993 *J. Comput. Phys.* **105** 279–89
[43] Herzog F, Ruijl B, Ueda T, Vermaseren J A M and Vogt A 2016 *PoS* **LL2016** 073 (*Preprint* 1608.01834)
[44] van Ritbergen T, Vermaseren J A M and Larin S A 1997 *Phys. Lett.* **B400** 379–84 (*Preprint* hep-ph/9701390)
[45] Czakon M 2005 *Nucl. Phys.* **B710** 485–98 (*Preprint* hep-ph/0411261)
[46] Ruijl B, Ueda T, Vermaseren J A M, Davies J and Vogt A 2016 *PoS* **LL2016** 071 (*Preprint* 1605.08408)
[47] Davies J, Vogt A, Ruijl B, Ueda T and Vermaseren J A M 2017 *Nucl. Phys.* **B915** 335–62 (*Preprint* 1610.07477)
[48] Lenstra A K, Lenstra H W and Lovász L 1982 *Math. Ann.* **261** 515–34
[49] Herzog F, Ruijl B, Ueda T, Vermaseren J A M and Vogt A 2017 *JHEP* **02** 090 (*Preprint* 1701.01404)
[50] Chetyrkin K G and Tkachov F V 1982 *Phys. Lett.* **B114** 340–4
[51] Chetyrkin K G and Smirnov V A 1984 *Phys. Lett.* **B144** 419–24
[52] Chetyrkin K G 2017 *Preprint* 1701.08627
[53] Herzog F and Ruijl B in preparation
[54] Baikov P A, Chetyrkin K G and Kühn J H 2017 *Phys. Rev. Lett.* **118** 082002 (*Preprint* 1606.08659)
[55] Luthe T, Maier A, Marquard P and Schröder Y 2016 *JHEP* **07** 127 (*Preprint* 1606.08662)
[56] Ruijl B, Ueda T, Vermaseren J A M and Vogt A in preparation
[57] Collins J C and Vermaseren J A M 2016 *Preprint* 1606.01177

Multi-loop calculations: numerical methods and applications

S. Borowka[1], **G. Heinrich**[2], **S. Jahn**[2], **S. P. Jones**[2], **M. Kerner**[2], **J. Schlenk**[3]

[1] Theoretical Physics Department, CERN, Geneva, Switzerland
[2] Max Planck Institute for Physics, Föhringer Ring 6, 80805 Munich, Germany
[3] IPPP, University of Durham, Durham DH1 3LE, UK

Abstract. We briefly review numerical methods for calculations beyond one loop and then describe new developments within the method of sector decomposition in more detail. We also discuss applications to two-loop integrals involving several mass scales.

1. Introduction

Precision calculations are of primary importance to scrutinise the Standard Model (SM) of particle physics and in particular the Higgs sector, where experiments moved from the discovery phase to the phase of precision measurements of the Higgs properties. Small deviations from the expected values may be our only hints to physics beyond the SM for some time, and therefore precise theoretical predictions are mandatory.

In the last decade, predictions at next-to-leading order (NLO) in perturbation theory in the strong coupling constant α_s got a large boost due to advances in calculational methods, and, together with NLO matching to parton shower Monte Carlo programs, became the state of the art to describe the data. However, for the phases II and III of the LHC, and even more so at future colliders, the situation is drastically different: the experimental precision for many important SM processes already has reached a level where NLO QCD predictions fall short. Therefore, a lot of effort has been spent in the past years to come up with corrections going beyond NLO QCD, ideally not only for total cross sections, but also for differential distributions.

A measure of complexity for the calculation of higher order corrections in perturbation theory involves the number of loops in the virtual amplitude, the number of scales (Mandelstam invariants, masses) and the number of external legs. While the problem of infrared subtractions is more severe the more massless particles are involved, the difficulty to obtain analytic expressions for master integrals at two loops and beyond increases rapidly as the number of mass scales grows. Therefore numerical methods to calculate loop integrals seem particularly well suited for integrals with several mass scales.

For processes involving only massless particles, virtual two-loop 4-point amplitudes have been calculated about 15 years ago [1–5]. After that, the main bottleneck to be overcome to achieve NNLO predictions for processes involving massless two-loop 4-point amplitudes was the lack of an efficient subtraction scheme for the infrared singularities occurring in the *real radiation part*, where up to two particles can become unresolved. Filling this gap has been a very active field of research in the past years. Various methods have been devised and are still under active development; they will be listed briefly below.

The next big step in the field of NNLO QCD corrections for $2 \to 2$ scattering processes was the availability of results for processes involving massive particles. Here the main problem currently resides in the *virtual two-loop part* of the calculation. Two problems are hampering progress here: (a) the *reduction* of the two-loop amplitudes to a minimal set of "master integrals" times coefficients gets increasingly complicated as the number of mass scales grows, and (b) the *analytic calculation* of the master integrals is extremely difficult, entering unexplored territory in terms of mathematical functions to express the occurring parameter integrals. Only very recently, analytic representations of two-loop integrals and amplitudes for $2 \to 2$ scattering processes involving massive particles became available, see e.g. [6–14].

2. Methods and tools for two-loop calculations and beyond

The steps to perform for the calculation of a (multi-)loop amplitude can be roughly divided into four stages: (1) generation of algebraic expressions for the amplitude, (2) reduction of the amplitude to a set of "master integrals" times coefficients, (3) isolation of the ultraviolet and infrared poles and (4) evaluation of the master integrals and combination with the coefficients to obtain the amplitude. To calculate a full cross section, loop amplitudes and real radiation contributions need to be combined, which requires a suitable scheme for the isolation of infrared-divergent real radiation, which is highly non-trivial beyond one loop. It also requires the construction of a stable and fast Monte Carlo program to perform the phase space integration.

It should be mentioned that stage (2) above is not mandatory. Reducing the set of integrals to a minimal "basis set" is usually beneficial, to reduce the number of integrals to calculate and to avoid large cancellations between linearly dependent integrals. However, it is also possible to evaluate the occurring integrals without reduction in a numerical approach, see e.g. [15–17]. Further, there are methods which aim to avoid the problems with IR singularities related to the split into real and virtual contributions by not performing such a partition at all [18–21].

In Table 1 we give a list of some publicly available multi-purpose tools which have been developed to perform the specific tasks described above, focusing on the numerical evaluation of the loop integrals. Certainly this list is incomplete and omits a multitude of codes which may be more efficient, but are tailored to more specific classes of integrals or amplitudes. Efforts towards the development of a package that can provide all the steps listed in Table 1 by combining QGRAF [22], FORM [23, 24], REDUZE [25, 26] and pySECDEC [27] are described in [28].

Table 2 shows some of the subtraction schemes for infrared divergent real radiation at NNLO.

Table 1. Public tools for various steps of loop amplitude calculations beyond one loop.

Step to be performed	available public tools
Diagram generation	QGRAF [22], FEYNARTS/FORMCALC [29, 30]
Amplitude manipulations	DIANA [31], FEYNCALC [32, 33]
Reduction	REDUZE [25, 26], FIRE [34, 35], LITERED [36, 37], AIR [38]
Numerical evaluation of the loop integrals	sector_decomposition [39], SECDEC [27, 40], FIESTA4 [41], NICODEMOS [42], AMBRE/MBNUMERICS [43, 44]

IOP Conf. Series: Journal of Physics: Conf. Series **920** (2017) 012003 doi:10.1088/1742-6596/920/1/012003

Concerning the reduction, we only listed the publicly available tools which, based on the integration-by-parts (IBP) method [45], can be used within a completely automated setup. Ideas how to reduce the computational complexity of IBP algorithms can be found in [46]. A fully automated system for amplitude generation and evaluation is also given by the GRACE system [47–49].

Novel reduction methods (see e.g. [50–61]), based on ideas such as integrand reduction and maximal cuts, are very promising, but have not reached the level of automation yet which is provided by the tools listed in Table 1.

Numerous methods for the numerical calculation of multi-loop integrals have been developed in addition to the ones mentioned above, we list only a few more recent ones here: direct numerical integration in momentum space [16], dispersion relations [62], use of the loop-tree duality [63], a toolbox of various dedicated numerical techniques [64, 65], numerical solution of differential equations [66, 67], numerical extrapolation method [68, 69], numerical evaluation of Mellin-Barnes integrals [70–72], private implementations of sector decomposition [64, 73–77].

Table 2. Methods for the isolation of IR divergent real radiation at NNLO.

method	analytic integration of subtraction terms	type/restrictions
antenna subtraction [78]	yes	subtraction
q_T subtraction [79]	yes	slicing; colourless final states
N-jettiness [80, 81]	yes	slicing
sector-improved residue subtraction [82–88]	no	subtraction
colourful subtraction [89, 90]	partly	subtraction; colourless initial states

3. Sector decomposition

Now we will describe the program SECDEC [27, 40, 91, 92] in more detail. The sector decomposition algorithm is described in [93, 94], which was inspired by earlier ideas as contained in [73, 95].

Higher order calculations in perturbation theory have in common that they involve multi-dimensional integrations over some parameters: Feynman (or Schwinger) parameters in the case of (multi-)loop integrals, or parameters related to the integration of subtraction terms over a factorised phase space in the case of infrared-divergent real radiation. Usually, these calculations are performed within the framework of dimensional regularisation, and one of the challenges is to factorise the poles in the regulator ϵ.

The program SECDEC [27, 40, 91, 92] is designed to perform this task in an automated way, and to integrate the coefficients of the resulting Laurent series in ϵ numerically.

The original sector decomposition algorithm described in Ref. [93] is based on an iterative procedure, which may run into an infinite recursion. It was pointed out however [39] that the structure of Feynman integrals is such that a decomposition algorithm must exist which is guaranteed to stop, as the procedure can be mapped to a known problem in convex geometry. In Ref. [96], an algorithm was presented which cannot lead to infinite recursion and is more efficient than previously employed algorithms with this property. SECDEC-3 and pySECDEC contain the implementation of a decomposition strategy (called G_2 in SECDEC-3 and `geometric` in pySECDEC), based on a modification of the method of Ref. [96], which usually outperforms the original iterative strategy (called X, or `iterative`).

IOP Conf. Series: Journal of Physics: Conf. Series **920** (2017) 012003 doi:10.1088/1742-6596/920/1/012003

3.1. Feynman parameter integrals

Multi-loop Feynman integrals can be written in a generic form. For ease of notation, we limit ourselves to scalar integrals here. Integrals with loop momenta in the numerator, or inverse propagators, only lead to an additional function in the numerator, and can be treated in the same way. We refer to [40, 94, 97] for further details.

A scalar Feynman integral G in D dimensions at L loops with N propagators, where the propagators can have arbitrary, not necessarily integer powers ν_j, has the following representation in momentum space:

$$G = \int \prod_{l=1}^{L} \mathrm{d}^D\kappa_l \frac{1}{\prod\limits_{j=1}^{N} P_j^{\nu_j}(\{k\},\{p\},m_j^2)} \tag{1}$$

$$\mathrm{d}^D\kappa_l = \frac{\mu^{4-D}}{i\pi^{\frac{D}{2}}}\,\mathrm{d}^D k_l\;,\; P_j(\{k\},\{p\},m_j^2) = q_j^2 - m_j^2 + i\delta\;,$$

where the q_j are linear combinations of external momenta p_i and loop momenta k_l.

Introducing Feynman parameters in Eq. (1) leads to

$$\begin{aligned} G &= \frac{\Gamma(N_\nu)}{\prod_{j=1}^{N}\Gamma(\nu_j)} \int_0^\infty \prod_{j=1}^{N} \mathrm{d}x_j\; x_j^{\nu_j-1}\,\delta(1-\sum_{i=1}^{N} x_i) \\ &\quad \cdot \int \mathrm{d}^D\kappa_1 \ldots \mathrm{d}^D\kappa_L \left[\sum_{i,j=1}^{L} k_i^{\mathrm{T}}\, M_{ij}\, k_j - 2\sum_{j=1}^{L} k_j^{\mathrm{T}} \cdot Q_j + J + i\,\delta\right]^{-N\nu} \\ &= \frac{(-1)^{N_\nu}}{\prod_{j=1}^{N}\Gamma(\nu_j)}\Gamma(N_\nu - LD/2) \int\limits_0^\infty \prod_{j=1}^{N} \mathrm{d}x_j\; x_j^{\nu_j-1}\,\delta(1-\sum_{l=1}^{N} x_l)\frac{\mathcal{U}^{N_\nu-(L+1)D/2}}{\mathcal{F}^{N_\nu-LD/2}}\;, \end{aligned} \tag{2}$$

where

$$\mathcal{F}(\vec{x}) = \det(M)\left[\sum_{j,l=1}^{L} Q_j\, M_{jl}^{-1}\, Q_l - J - i\,\delta\right]\;, \tag{3}$$

$$\mathcal{U}(\vec{x}) = \det(M),\;\; N_\nu = \sum_{j=1}^{N} \nu_j\;. \tag{4}$$

In the expressions above, M is an $L \times L$ matrix containing Feynman parameters, Q is an L-dimensional vector, where each entry is a linear combination of external momenta and Feynman parameters, and J is a scalar expression containing kinematic invariants and Feynman parameters.

$\mathcal{U}$ is a positive semi-definite function, which vanishes at the UV subdivergences of the graph. In the region where all invariants formed from external momenta are negative ("*Euclidean region*"), $\mathcal{F}$ is also a positive semi-definite function of the Feynman parameters x_j. If some of the invariants are zero, for example if some of the external momenta are light-like, an IR divergence may appear and $\mathcal{F}$ vanishes for certain points in parameter space. In the Euclidean region, the necessary condition $\mathcal{F} = 0$ for an IR divergence can only be fulfilled if some of the parameters x_i are zero. The endpoint singularities of both UV and IR nature can be regulated by dimensional regularisation and factored out of the functions $\mathcal{U}$ and $\mathcal{F}$ using sector decomposition.

The basic concept of sector decomposition is the following: We consider a two-dimensional parameter integral which contains a singular region where both x and y vanish:

$$I = \int_0^1 dx \int_0^1 dy\, x^{-1-a\epsilon}\, y^{-b\epsilon}\, (x+y)^{-1}\,. \tag{5}$$

Our aim is to factorise the singularities for $x \to 0$ and $y \to 0$. Therefore we divide the integration range into two sectors where x and y are ordered:

$$I = \int_0^1 dx \int_0^1 dy\, x^{-1-a\epsilon}\, y^{-b\epsilon}\, (x+y)^{-1} \, [\underbrace{\Theta(x-y)}_{(1)} + \underbrace{\Theta(y-x)}_{(2)}]\,.$$

Now we substitute $y = x\,t$ in sector (1) and $x = y\,t$ in sector (2) to remap the integration range to the unit square and obtain

$$I = \int_0^1 dx\, x^{-1-(a+b)\epsilon} \int_0^1 dt\, t^{-b\epsilon}\, (1+t)^{-1} + \int_0^1 dy\, y^{-1-(a+b)\epsilon} \int_0^1 dt\, t^{-1-a\epsilon}\, (1+t)^{-1}\,. \tag{6}$$

This way the singularities are factorised into monomials, while the remaining denominator goes to a constant if the integration variables approach zero. For more complicated integrands, this procedure can be iterated until a complete factorisation is achieved.

However, after the UV and IR singularities have been extracted as poles in $1/\epsilon$, for non-Euclidean kinematics integrable singularities related to kinematic thresholds remain. These singularities imply that $\mathcal{F}$ is vanishing inside the integration region for some combinations of Feynman parameter values and values of the kinematic invariants. However, the integrals can be evaluated by deforming the integration contour into the complex plane [18], as explained in detail in Refs. [92, 98].

3.2. Program structure

The program consists of two main parts, an algebraic and a numerical part. The algebraic part constructs the integrand from the list of propagators or from the graph labels, performs the sector decomposition procedure to factorise the poles in the regulator ϵ, the subtractions and the expansion in ϵ, and prepares the contour deformation in the case of non-Euclidean kinematics. In SecDec-3, all the algebraic steps are performed in `Mathematica`. In the new version [27], the algebraic part has been completely restructured and implemented in `python`, therefore the new version is called pySecDec.

The numerical part consists of C++ functions which are integrated numerically with the Cuba library [99]. The new program pySecDec produces C++ code using `Form` [23, 24], and in addition produces C++ libraries such that the finite parametric functions representing an integral after the algebraic procedure can be linked to other programs. The basic workflow is shown in Figs. 1 and 2.

3.3. Recent program developments

In addition to the new possibilities of usage, there are various new features in pySecDec compared to SecDec-3.0:

- the functions can have any number of different regulators, not only the dimensional regulator ϵ, needed for example in analytic regularisation within Soft-Collinear Effective Theory [100];
- numerators of loop integrals can be defined in terms of contracted Lorentz vectors or inverse propagators or a combination of both;
- the distinction between "general functions" and "loop integrands" is removed in the sense that all features which are not loop-integral-specific are also available for general polynomial functions;

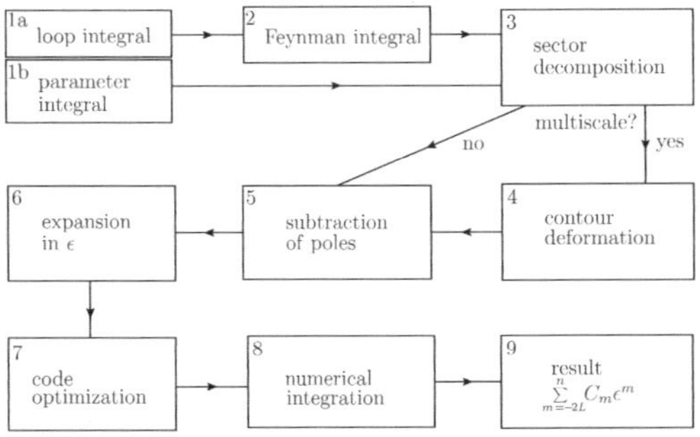

Figure 1. Flowchart showing the main steps the program performs to produce the numerical result as a Laurent series in ϵ. L denotes the number of loops.

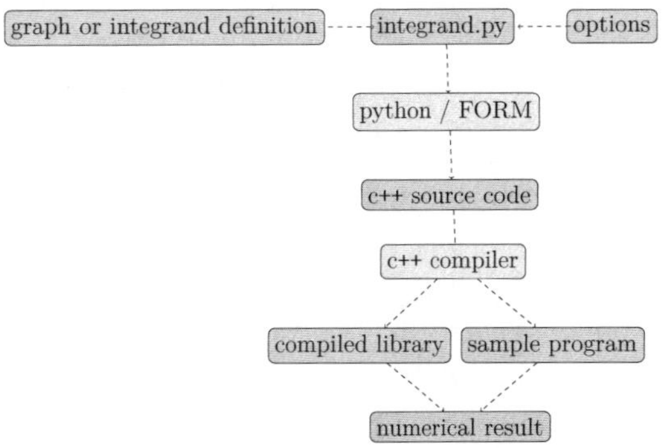

Figure 2. Basic workflow of pySecDec .

- the inclusion of "user-defined" functions which do not enter the decomposition has been facilitated and extended;
- the treatment of poles which are higher than logarithmic has been improved;
- a procedure has been implemented to detect and remap spurious singularities which cannot be cured by contour deformation;
- a symmetry finder has been added which can detect possible isomorphisms between sectors.

Version 1 of pySecDec [27] is available at `http://secdec.hepforge.org/`.

3.4. Phenomenological application

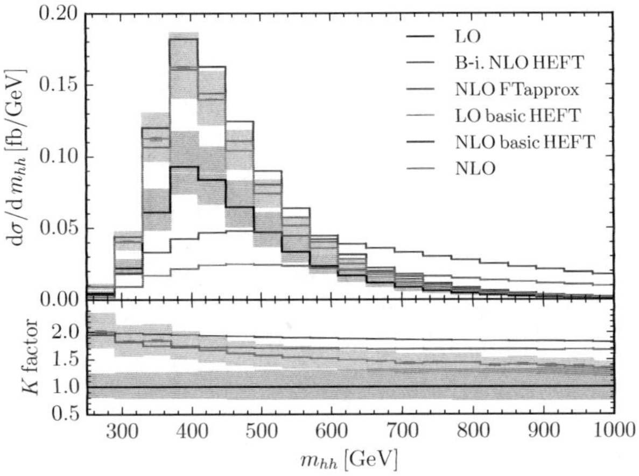

Figure 3. Higgs boson pair invariant mass distribution with full top quark mass dependence compared to various approximations. B-i. NLO HEFT denotes the Born-improved HEFT approximation, while "basic HEFT" is without the rescaling by the full Born level result. "FTapprox" stands for an approximation where the real radiation part is calculated with full mass dependence, while the virtual part is given by the Born-improved HEFT approximation.

The numerical approach based on SecDec has been applied to calculate massive two-loop integrals entering $gg \to HH$ at NLO, retaining the full top quark mass dependence [17,101,102]. The calculation is based on the setup described in Refs. [28,101,103]. The amplitude generation leads to about 10000 integrals before any symmetries are taken into account, which have been reduced to $\mathcal{O}(300)$ integrals using Reduze [25,26]. A complete reduction could not be obtained for the non-planar 4-point integrals. The inverse propagators appearing in unreduced integrals were rewritten in terms of scalar products and directly computed with SecDec.

For the total cross section at $\sqrt{s} = 14\,\text{TeV}$, we found a reduction of about 14% when including the full top quark mass dependence as compared to the Born-improved HEFT ("Higgs Effective Field Theory") approximation, where in the latter the NLO corrections are calculated in the $m_t \to \infty$ limit, and "Born-improved" means that the result obtained in the $m_t \to \infty$ limit is rescaled with the full Born level result divided by the HEFT Born level result. Fig. 3 shows results for the Higgs boson pair invariant mass distribution. For further details we refer to [17, 101, 102].

4. Conclusions

We have given a brief overview on numerical methods to calculate integrals (and cross sections) beyond one-loop order, before focusing on the program SecDec, in particular the new version pySecDec. We pointed to its application within a context that goes beyond the calculation of individual master integrals, for example the possibility to use it as a library to evaluate two-loop amplitudes where the analytic expressions for the master integrals are not known.

Acknowledgments

G.H. would like to thank the organizers of CPP2016 for the great workshop, and dedicate these proceedings to the memory of Shimizu-Sensei. We also would like to thank Nicolas Greiner and Tom Zirke for collaboration on parts of the projects presented here. This research was supported in part by the Research Executive Agency (REA) of the European Union under the Grant Agreement PITN-GA2012316704 (HiggsTools) and the ERC Advanced Grant MC@NNLO (340983). S. Borowka gratefully acknowledges financial support by the ERC Starting Grant "MathAm" (39568).

References

[1] E. W. N. Glover, C. Oleari and M. E. Tejeda-Yeomans, *Two loop QCD corrections to gluon-gluon scattering*, *Nucl. Phys.* **B605** (2001) 467–485, [hep-ph/0102201].

[2] L. W. Garland, T. Gehrmann, E. W. N. Glover, A. Koukoutsakis and E. Remiddi, *The Two loop QCD matrix element for $e^+e^- \to 3$ jets*, *Nucl. Phys.* **B627** (2002) 107–188, [hep-ph/0112081].

[3] Z. Bern, A. De Freitas and L. J. Dixon, *Two loop helicity amplitudes for gluon-gluon scattering in QCD and supersymmetric Yang-Mills theory*, *JHEP* **03** (2002) 018, [hep-ph/0201161].

[4] S. Moch, P. Uwer and S. Weinzierl, *Two loop amplitudes with nested sums: Fermionic contributions to $e^+e^- \to q\bar{q}g$*, *Phys. Rev.* **D66** (2002) 114001, [hep-ph/0207043].

[5] T. Binoth, E. W. N. Glover, P. Marquard and J. J. van der Bij, *Two loop corrections to light by light scattering in supersymmetric QED*, *JHEP* **05** (2002) 060, [hep-ph/0202266].

[6] J. M. Henn, *Multiloop integrals in dimensional regularization made simple*, *Phys. Rev. Lett.* **110** (2013) 251601, [1304.1806].

[7] J. M. Henn, K. Melnikov and V. A. Smirnov, *Two-loop planar master integrals for the production of off-shell vector bosons in hadron collisions*, *JHEP* **05** (2014) 090, [1402.7078].

[8] F. Caola, J. M. Henn, K. Melnikov and V. A. Smirnov, *Non-planar master integrals for the production of two off-shell vector bosons in collisions of massless partons*, *JHEP* **09** (2014) 043, [1404.5590].

[9] T. Gehrmann, A. von Manteuffel and L. Tancredi, *The two-loop helicity amplitudes for $q\bar{q}' \to V_1V_2 \to 4$ leptons*, *JHEP* **09** (2015) 128, [1503.04812].

[10] A. von Manteuffel and L. Tancredi, *The two-loop helicity amplitudes for $gg \to V_1V_2 \to 4$* leptons, *JHEP* **06** (2015) 197, [1503.08835].

[11] F. Caola, J. M. Henn, K. Melnikov, A. V. Smirnov and V. A. Smirnov, *Two-loop helicity amplitudes for the production of two off-shell electroweak bosons in gluon fusion*, *JHEP* **06** (2015) 129, [1503.08759].

[12] R. Bonciani, S. Di Vita, P. Mastrolia and U. Schubert, *Two-Loop Master Integrals for the mixed EW-QCD virtual corrections to Drell-Yan scattering*, *JHEP* **09** (2016) 091, [1604.08581].

[13] R. Bonciani, V. Del Duca, H. Frellesvig, J. M. Henn, F. Moriello and V. A. Smirnov, *Two-loop planar master integrals for Higgs $\to 3$ partons with full heavy-quark mass dependence*, *JHEP* **12** (2016) 096, [1609.06685].

[14] A. Primo and L. Tancredi, *On the maximal cut of Feynman integrals and the solution of their differential equations*, *Nucl. Phys.* **B916** (2017) 94–116, [1610.08397].

[15] M. Spira, A. Djouadi, D. Graudenz and P. M. Zerwas, *Higgs boson production at the LHC*, *Nucl. Phys.* **B453** (1995) 17–82, [hep-ph/9504378].

[16] S. Becker and S. Weinzierl, *Direct numerical integration for multi-loop integrals*, *Eur. Phys. J.* **C73** (2013) 2321, [1211.0509].

[17] S. Borowka, N. Greiner, G. Heinrich, S. Jones, M. Kerner, J. Schlenk et al., *Higgs Boson Pair Production in Gluon Fusion at Next-to-Leading Order with Full Top-Quark Mass Dependence*, *Phys. Rev. Lett.* **117** (2016) 012001, [1604.06447].

[18] D. E. Soper, *Techniques for QCD calculations by numerical integration*, *Phys. Rev.* **D62** (2000) 014009, [hep-ph/9910292].

[19] D. A. Forde and A. Signer, *Infrared finite amplitudes for massless gauge theories*, *Nucl. Phys.* **B684** (2004) 125–161, [hep-ph/0311059].

[20] S. Catani, T. Gleisberg, F. Krauss, G. Rodrigo and J.-C. Winter, *From loops to trees by-passing Feynman's theorem*, *JHEP* **09** (2008) 065, [0804.3170].

[21] G. F. R. Sborlini, F. Driencourt-Mangin and G. Rodrigo, *Four-dimensional unsubtraction with massive particles*, *JHEP* **10** (2016) 162, [1608.01584].

[22] P. Nogueira, *Automatic Feynman graph generation*, *J.Comput.Phys.* **105** (1993) 279–289.

[23] J. Vermaseren, *New features of FORM*, math-ph/0010025.

[24] J. Kuipers, T. Ueda and J. A. M. Vermaseren, *Code Optimization in FORM*, *Comput. Phys. Commun.* **189** (2015) 1–19, [1310.7007].

[25] C. Studerus, *Reduze-Feynman Integral Reduction in C++*, *Comput. Phys. Commun.* **181** (2010) 1293–1300, [0912.2546].

[26] A. von Manteuffel and C. Studerus, *Reduze 2 - Distributed Feynman Integral Reduction*, 1201.4330.

[27] S. Borowka, G. Heinrich, S. Jahn, S. P. Jones, M. Kerner, J. Schlenk et al., *pySecDec: a toolbox for the numerical evaluation of multi-scale integrals*, 1703.09692.

[28] S. P. Jones, *Automation of 2-loop Amplitude Calculations*, *PoS* **LL2016** (2016) 069, [1608.03846].

[29] T. Hahn, *Generating Feynman diagrams and amplitudes with FeynArts 3*, *Comput. Phys. Commun.* **140** (2001) 418–431, [hep-ph/0012260].

[30] T. Hahn, S. Paehr and C. Schappacher, *FormCalc 9 and Extensions*, *PoS* **LL2016** (2016) 068, [1604.04611].

[31] M. Tentyukov and J. Fleischer, *A Feynman diagram analyzer DIANA*, *Comput. Phys. Commun.* **132** (2000) 124–141, [hep-ph/9904258].

[32] V. Shtabovenko, R. Mertig and F. Orellana, *New Developments in FeynCalc 9.0*, *Comput. Phys. Commun.* **207** (2016) 432–444, [1601.01167].

[33] V. Shtabovenko, *FeynHelpers: Connecting FeynCalc to FIRE and Package-X*, 1611.06793.

[34] A. V. Smirnov, *Algorithm FIRE – Feynman Integral REduction*, *JHEP* **10** (2008) 107, [0807.3243].

[35] A. V. Smirnov, *FIRE5: a C++ implementation of Feynman Integral REduction*, *Comput. Phys. Commun.* **189** (2014) 182–191, [1408.2372].

[36] R. N. Lee, *Presenting LiteRed: a tool for the Loop InTEgrals REDuction*, 1212.2685.

[37] R. N. Lee, *LiteRed 1.4: a powerful tool for reduction of multiloop integrals*, *J. Phys. Conf. Ser.* **523** (2014) 012059, [1310.1145].

[38] C. Anastasiou and A. Lazopoulos, *Automatic integral reduction for higher order perturbative calculations*, *JHEP* **07** (2004) 046, [hep-ph/0404258].

[39] C. Bogner and S. Weinzierl, *Resolution of singularities for multi-loop integrals*, *Comput.Phys.Commun.* **178** (2008) 596–610, [0709.4092].

[40] S. Borowka, G. Heinrich, S. P. Jones, M. Kerner, J. Schlenk and T. Zirke, *SecDec-3.0: numerical evaluation of multi-scale integrals beyond one loop*, *Comput. Phys. Commun.* **196** (2015) 470–491, [1502.06595].

[41] A. V. Smirnov, *FIESTA4: Optimized Feynman integral calculations with GPU support*, *Comput. Phys. Commun.* **204** (2016) 189–199, [1511.03614].

[42] A. Freitas, *Numerical evaluation of multi-loop integrals using subtraction terms*, *JHEP* **07** (2012) 132, [1205.3515].

[43] J. Gluza, K. Kajda, T. Riemann and V. Yundin, *Numerical Evaluation of Tensor Feynman Integrals in Euclidean Kinematics*, *Eur.Phys.J.* **C71** (2011) 1516, [1010.1667].

[44] I. Dubovyk, J. Gluza, T. Riemann and J. Usovitsch, *Numerical integration of massive two-loop Mellin-Barnes integrals in Minkowskian regions*, *PoS* **LL2016** (2016) 034, [1607.07538].

[45] S. Laporta, *High precision calculation of multiloop Feynman integrals by difference equations*, *Int. J. Mod. Phys.* **A15** (2000) 5087–5159, [hep-ph/0102033].

[46] A. von Manteuffel and R. M. Schabinger, *A novel approach to integration by parts reduction*, *Phys. Lett.* **B744** (2015) 101–104, [1406.4513].

[47] G. Belanger, F. Boudjema, J. Fujimoto, T. Ishikawa, T. Kaneko, K. Kato et al., *Automatic calculations in high energy physics and Grace at one-loop*, *Phys. Rept.* **430** (2006) 117–209, [hep-ph/0308080].
[48] J. Fujimoto et al., *The GRACE project: QCD, SUSY, multi-loop*, *PoS* **RADCOR2011** (2011) 012.
[49] P. H. Khiem et al., *Full $\mathcal{O}(\alpha)$ electroweak radiative corrections to $e^+e^- \to e^+e^-\gamma$ at the ILC with GRACE-Loop*, *Phys. Lett.* **B740** (2015) 192–198, [1403.6557].
[50] D. A. Kosower and K. J. Larsen, *Maximal Unitarity at Two Loops*, *Phys. Rev.* **D85** (2012) 045017, [1108.1180].
[51] P. Mastrolia, E. Mirabella, G. Ossola and T. Peraro, *Integrand-Reduction for Two-Loop Scattering Amplitudes through Multivariate Polynomial Division*, *Phys. Rev.* **D87** (2013) 085026, [1209.4319].
[52] S. Badger, H. Frellesvig and Y. Zhang, *Hepta-Cuts of Two-Loop Scattering Amplitudes*, *JHEP* **04** (2012) 055, [1202.2019].
[53] P. Mastrolia, E. Mirabella, G. Ossola and T. Peraro, *Multiloop Integrand Reduction for Dimensionally Regulated Amplitudes*, *Phys. Lett.* **B727** (2013) 532–535, [1307.5832].
[54] K. J. Larsen and Y. Zhang, *Integration-by-parts reductions from unitarity cuts and algebraic geometry*, *Phys. Rev.* **D93** (2016) 041701, [1511.01071].
[55] H. Ita, *Two-loop Integrand Decomposition into Master Integrals and Surface Terms*, *Phys. Rev.* **D94** (2016) 116015, [1510.05626].
[56] S. Badger, G. Mogull and T. Peraro, *Local integrands for two-loop all-plus Yang-Mills amplitudes*, *JHEP* **08** (2016) 063, [1606.02244].
[57] P. Mastrolia, T. Peraro and A. Primo, *Adaptive Integrand Decomposition in parallel and orthogonal space*, *JHEP* **08** (2016) 164, [1605.03157].
[58] A. Georgoudis, K. J. Larsen and Y. Zhang, *Azurite: An algebraic geometry based package for finding bases of loop integrals*, 1612.04252.
[59] T. Peraro, *Scattering amplitudes over finite fields and multivariate functional reconstruction*, *JHEP* **12** (2016) 030, [1608.01902].
[60] S. Abreu, F. Febres Cordero, H. Ita, M. Jaquier and B. Page, *Subleading Poles in the Numerical Unitarity Method at Two Loops*, 1703.05255.
[61] S. Abreu, F. Febres Cordero, H. Ita, M. Jaquier, B. Page and M. Zeng, *Two-Loop Four-Gluon Amplitudes with the Numerical Unitarity Method*, 1703.05273.
[62] S. Bauberger and A. Freitas, *TVID: Three-loop Vacuum Integrals from Dispersion relations*, 1702.02996.
[63] G. Chachamis, S. Buchta, P. Draggiotis and G. Rodrigo, *Attacking One-loop Multi-leg Feynman Integrals with the Loop-Tree Duality*, *PoS* **DIS2016** (2016) 067, [1607.00875].
[64] G. Passarino and S. Uccirati, *Two-loop vertices in quantum field theory: Infrared and collinear divergent configurations*, *Nucl.Phys.* **B747** (2006) 113–189, [hep-ph/0603121].
[65] S. Actis, G. Passarino, C. Sturm and S. Uccirati, *NNLO Computational Techniques: The Cases $H \to \gamma\gamma$ and $H \to gg$*, *Nucl. Phys.* **B811** (2009) 182–273, [0809.3667].
[66] P. Bärnreuther, M. Czakon and P. Fiedler, *Virtual amplitudes and threshold behaviour of hadronic top-quark pair-production cross sections*, *JHEP* **02** (2014) 078, [1312.6279].
[67] M. Czakon, *Tops from Light Quarks: Full Mass Dependence at Two-Loops in QCD*, *Phys. Lett.* **B664** (2008) 307–314, [0803.1400].
[68] E. de Doncker, F. Yuasa, K. Kato, T. Ishikawa, J. Kapenga and O. Olagbemi, *Regularization with Numerical Extrapolation for Finite and UV-Divergent Multi-loop Integrals*, 1702.04904.
[69] K. Kato, E. de Doncker, T. Ishikawa, J. Kapenga, O. Olagbemi and F. Yuasa, *High Performance and Increased Precision Techniques for Feynman Loop Integrals*, *J. Phys. Conf. Ser.* **762** (2016) 012070.
[70] M. Czakon, *Automatized analytic continuation of Mellin-Barnes integrals*, *Comput. Phys. Commun.* **175** (2006) 559–571, [hep-ph/0511200].
[71] J. Gluza, T. Jelinski and D. A. Kosower, *Efficient Evaluation of Massive Mellin–Barnes Integrals*, 1609.09111.
[72] I. Dubovyk, A. Freitas, J. Gluza, T. Riemann and J. Usovitsch, *The two-loop electroweak bosonic corrections to $\sin^2\theta^{\rm b}_{\rm eff}$*, *Phys. Lett.* **B762** (2016) 184–189, [1607.08375].
[73] M. Roth and A. Denner, *High-energy approximation of one-loop Feynman integrals*, *Nucl. Phys.* **B479** (1996) 495–514, [hep-ph/9605420].
[74] A. Denner and S. Pozzorini, *An Algorithm for the high-energy expansion of multi-loop diagrams to next-to-leading logarithmic accuracy*, *Nucl. Phys.* **B717** (2005) 48–85, [hep-ph/0408068].
[75] C. Anastasiou, S. Beerli and A. Daleo, *Evaluating multi-loop Feynman diagrams with infrared and threshold singularities numerically*, *JHEP* **05** (2007) 071, [hep-ph/0703282].
[76] T. Ueda and J. Fujimoto, *New implementation of the sector decomposition on FORM*, *PoS* **ACAT08** (2008) 120, [0902.2656].
[77] C. Anastasiou, F. Herzog and A. Lazopoulos, *On the factorization of overlapping singularities at NNLO*,

JHEP **03** (2011) 038, [1011.4867].
[78] A. Gehrmann-De Ridder, T. Gehrmann and E. W. N. Glover, *Antenna subtraction at NNLO*, *JHEP* **09** (2005) 056, [hep-ph/0505111].
[79] S. Catani, L. Cieri, G. Ferrera, D. de Florian and M. Grazzini, *Vector boson production at hadron colliders: a fully exclusive QCD calculation at NNLO*, *Phys. Rev. Lett.* **103** (2009) 082001, [0903.2120].
[80] J. Gaunt, M. Stahlhofen, F. J. Tackmann and J. R. Walsh, *N-jettiness Subtractions for NNLO QCD Calculations*, *JHEP* **09** (2015) 058, [1505.04794].
[81] R. Boughezal, C. Focke, X. Liu and F. Petriello, *W-boson production in association with a jet at next-to-next-to-leading order in perturbative QCD*, *Phys. Rev. Lett.* **115** (2015) 062002, [1504.02131].
[82] G. Heinrich, *A numerical method for NNLO calculations*, *Nucl. Phys. Proc. Suppl.* **116** (2003) 368–372, [hep-ph/0211144].
[83] C. Anastasiou, K. Melnikov and F. Petriello, *A new method for real radiation at NNLO*, *Phys. Rev.* **D69** (2004) 076010, [hep-ph/0311311].
[84] T. Binoth and G. Heinrich, *Numerical evaluation of phase space integrals by sector decomposition*, *Nucl. Phys.* **B693** (2004) 134–148, [hep-ph/0402265].
[85] M. Czakon, *A novel subtraction scheme for double-real radiation at NNLO*, *Phys. Lett.* **B693** (2010) 259–268, [1005.0274].
[86] R. Boughezal, K. Melnikov and F. Petriello, *A subtraction scheme for NNLO computations*, *Phys. Rev.* **D85** (2012) 034025, [1111.7041].
[87] M. Czakon and D. Heymes, *Four-dimensional formulation of the sector-improved residue subtraction scheme*, *Nucl. Phys.* **B890** (2014) 152–227, [1408.2500].
[88] F. Caola, K. Melnikov and R. Röntsch, *Nested soft-collinear subtractions in NNLO QCD computations*, *Eur. Phys. J.* **C77** (2017) 248, [1702.01352].
[89] G. Somogyi, Z. Trocsanyi and V. Del Duca, *A Subtraction scheme for computing QCD jet cross sections at NNLO: Regularization of doubly-real emissions*, *JHEP* **01** (2007) 070, [hep-ph/0609042].
[90] V. Del Duca, C. Duhr, A. Kardos, G. Somogyi and Z. Trcsnyi, *Three-Jet Production in Electron-Positron Collisions at Next-to-Next-to-Leading Order Accuracy*, *Phys. Rev. Lett.* **117** (2016) 152004, [1603.08927].
[91] J. Carter and G. Heinrich, *SecDec: A general program for sector decomposition*, *Comput. Phys. Commun.* **182** (2011) 1566–1581, [1011.5493].
[92] S. Borowka, J. Carter and G. Heinrich, *Numerical Evaluation of Multi-Loop Integrals for Arbitrary Kinematics with SecDec 2.0*, *Comput. Phys. Commun.* **184** (2013) 396–408, [1204.4152].
[93] T. Binoth and G. Heinrich, *An automatized algorithm to compute infrared divergent multiloop integrals*, *Nucl. Phys.* **B585** (2000) 741–759, [hep-ph/0004013].
[94] G. Heinrich, *Sector Decomposition*, *Int. J. Mod. Phys.* **A23** (2008) 1457–1486, [0803.4177].
[95] K. Hepp, *Proof of the Bogolyubov-Parasiuk theorem on renormalization*, *Commun. Math. Phys.* **2** (1966) 301–326.
[96] T. Kaneko and T. Ueda, *A Geometric method of sector decomposition*, *Comput.Phys.Commun.* **181** (2010) 1352–1361, [0908.2897].
[97] J. Schlenk and T. Zirke, *Calculation of Multi-Loop Integrals with SecDec-3.0*, in *Proceedings, 12th International Symposium on Radiative Corrections (Radcor 2015) and LoopFest XIV (Radiative Corrections for the LHC and Future Colliders): Los Angeles, CA, USA, June 15-19, 2015*, 2016. 1601.03982.
[98] S. C. Borowka, *Evaluation of multi-loop multi-scale integrals and phenomenological two-loop applications*. PhD thesis, Munich, Tech. U., 2014. 1410.7939.
[99] T. Hahn, *CUBA: A Library for multidimensional numerical integration*, *Comput. Phys. Commun.* **168** (2005) 78–95, [hep-ph/0404043].
[100] T. Becher and G. Bell, *Analytic Regularization in Soft-Collinear Effective Theory*, *Phys. Lett.* **B713** (2012) 41–46, [1112.3907].
[101] S. Borowka, N. Greiner, G. Heinrich, S. P. Jones, M. Kerner, J. Schlenk et al., *Full top quark mass dependence in Higgs boson pair production at NLO*, *JHEP* **10** (2016) 107, [1608.04798].
[102] G. Heinrich, S. P. Jones, M. Kerner, G. Luisoni and E. Vryonidou, *NLO predictions for Higgs boson pair production with full top quark mass dependence matched to parton showers*, 1703.09252.
[103] M. Kerner, *Next-to-Leading Order Corrections to Higgs Boson Pair Production in Gluon Fusion*, *PoS* **LL2016** (2016) 023, [1608.03851].

IOP Conf. Series: Journal of Physics: Conf. Series **920** (2017) 012004 doi:10.1088/1742-6596/920/1/012004

Methods for deriving functional equations for Feynman integrals

O. V. Tarasov

Joint Institute for Nuclear Research, 141980 Dubna (Moscow Region), Russia.

Abstract. We present short review of two methods for obtaining functional equations for Feynman integrals. Application of these methods for finding functional equations for one- and two- loop integrals described in detail. It is shown that with the aid of functional equations Feynman integrals in general kinematics can be expressed in terms of simpler integrals. Similarities between Feynman functional equations for Feynman integrals and addition theorem for Abel integrals are shortly discussed.

PACS numbers: 02.30.Gp, 02.30.Ks, 12.20.Ds, 12.38.Bx
Keywords: Feynman integrals, functional equations

1. Introduction

Functional equations (FE) for Feynman integrals for the first time were proposed in Ref. [1]. Details of derivation of such equations for one-loop integrals were described in Refs. [2], [3]. In Ref.[4] a new method for derivation of functional equations was proposed. This method is based on finding algebraic relations between products of propagators. By integrating different combinations of these algebraic relations one can easily obtain functional equations for multiloop integrals. Our functional relations can be used for expressing integrals with nontrivial kinematical dependence in terms of much simpler integrals. Also functional equations can be exploited for analytic continuation of Feynman integrals into different kinematical domains. In the next sections we will explain how to derive FE and will present several illustrative examples.

2. Derivation of FE from recurrence relations

The method proposed in Ref.[1] is based on the use of recurrence relations between integrals. The most general recurrence relation for Feynman integrals can be written as

$$\sum_j Q_j I_{j,n} = \sum_{k,r<n} R_{k,r} I_{k,r}, \tag{1}$$

where Q_j, R_k are polynomials in masses, scalar products of external momenta, space-time dimension d, and powers of propagators. $I_{k,r}$ - are integrals with r external lines. The index k just labels different integrals in a recurrence relation. In recurrence relations some integrals are more complicated than the others. The complexity of the integral depends on the number of their arguments and number of loops. The main idea of deriving functional equation from a recurrence relation can be formulated as follows. By choosing kinematical variables, masses, indices of propagators remove most complicated integrals from the relation, i.e. impose conditions :

$$Q_j = 0, \tag{2}$$

keeping at least some other coefficients $R_k \neq 0$.

To demonstrate how this idea works in practice let's consider one-loop n-point integrals $I_n^{(d)}$. These integrals satisfy the so-called generalized recurrence relations [5]:

$$G_{n-1}\nu_j \mathbf{j}^+ I_n^{(d+2)} - (\partial_j \Delta_n) I_n^{(d)} = \sum_{k=1}^{n} (\partial_j \partial_k \Delta_n) \mathbf{k}^- I_n^{(d)}, \tag{3}$$

where $\mathbf{j}^\pm$ shifts indices $\nu_j \to \nu_j \pm 1$, $\partial_j \equiv \frac{\partial}{\partial m_j^2}$,

$$G_{n-1} = -2^n \begin{vmatrix} p_1p_1 & p_1p_2 & \dots & p_1p_{n-1} \\ \vdots & \vdots & \ddots & \vdots \\ p_1p_{n-1} & p_2p_{n-1} & \dots & p_{n-1}p_{n-1} \end{vmatrix}, \tag{4}$$

$$\Delta_n = \begin{vmatrix} Y_{11} & Y_{12} & \dots & Y_{1n} \\ \vdots & \vdots & \ddots & \vdots \\ Y_{1n} & Y_{2n} & \dots & Y_{nn} \end{vmatrix}, \qquad Y_{ij} = m_i^2 + m_j^2 - p_{ij}, \qquad p_{ij} = (p_i - p_j)^2. \tag{5}$$

2.1. One-loop propagator type integral

Setting $n = 3$, $j = 1$, $m_3^2 = 0$ in Eq.(3) leads to an expression connecting integrals $I_3^{(d+2)}$, $I_3^{(d)}$ and $I_2^{(d)}$. In order to get rid of integrals $I_3^{(d+2)}$, $I_3^{(d)}$ we must impose conditions:

$$G_2 = 0, \qquad \Delta_3 = 0. \tag{6}$$

Solving this system for p_{13}, p_{23} yields FE for the integral $I_2^{(d)}$:

$$\begin{aligned} I_2^{(d)}(m_1^2, m_2^2, p_{12}) &= \frac{p_{12} + m_1^2 - m_2^2 - \alpha_{12}}{2p_{12}} I_2^{(d)}(m_1^2, 0, s_{13}) \\ &+ \frac{p_{12} - m_1^2 + m_2^2 + \alpha_{12}}{2p_{12}} I_2^{(d)}(0, m_2^2, s_{23}), \end{aligned}$$

where

$$\begin{aligned} s_{13} &= \frac{\Delta_{12} + 2p_{12}m_1^2 - (p_{12} + m_1^2 - m_2^2)\alpha_{12}}{2p_{12}}, \\ s_{23} &= \frac{\Delta_{12} + 2p_{12}m_2^2 + (p_{12} - m_1^2 + m_2^2)\alpha_{12}}{2p_{12}}, \\ \alpha_{12} &= \pm\sqrt{\Delta_{12}}. \\ \Delta_{ij} &= p_{ij}^2 + m_i^4 + m_j^4 - 2p_{ij}m_i^2 - 2p_{ij}m_j^2 - 2m_i^2m_j^2. \end{aligned}$$

This equation represent integral with arbitrary masses m_1, m_2 and arbitrary external momentum in terms of integrals with one propagator massless. $I_2^{(d)}$ integrals with one massless propagator can be expressed in terms of Gauss' hypergeometric function ${}_2F_1$ [6], [7]. Evaluation of such integrals is much easier than evaluation of integrals with two masses.

2.2. FE for one-loop vertex type integral

At $n = 4$, $j = 1$, $m_4 = 0$, in order to get rid of integrals $I_4^{(d)}$, $I_4^{(d+2)}$ and also one vertex integral $I_3^{(d)}$ in Eq.(3), we impose conditions

$$G_3 = 0, \qquad \partial_1 \Delta_4 = 0, \qquad \partial_1 \partial_2 \Delta_4 = 0. \tag{7}$$

This system of equations depends on 10 variables $p_{12}, p_{13}, p_{14}, p_{23}, p_{24}, p_{34}$, $m_1^2, m_2^2, m_3^2, m_4^2$ and it can be solved by excluding, for example, p_{14}, p_{24} and p_{34}. Substituting solution for p_{14}, p_{24}, p_{34} into Eq.(3) leads to the following FE:

$$\begin{aligned}
&I_3^{(d)}(m_1^2, m_2^2, m_3^2, s_{23}, s_{13}, s_{12}) = \\
&\frac{s_{13} + m_3^2 - m_1^2 + \alpha_{13}}{2s_{13}} I_3^{(d)}(m_2^2, m_3^2, 0, s_{34}^{(13)}, s_{24}(m_1^2, m_3^2, s_{23}, s_{13}, s_{12}), s_{23}) \\
&+ \frac{s_{13} - m_3^2 + m_1^2 - \alpha_{13}}{2s_{13}} I_3^{(d)}(m_1^2, m_2^2, 0, s_{24}(m_1^2, m_3^2, s_{23}, s_{13}, s_{12}), s_{14}^{(13)}, s_{12}),
\end{aligned} \tag{8}$$

where

$$\begin{aligned}
p_{14} &= s_{14}^{(13)}, \\
p_{24} &= s_{24}(m_1^2, m_3^2, p_{23}, p_{13}, p_{12}) \\
&= \frac{(p_{12} + p_{23} - m_1^2 - m_3^2)p_{13} + (p_{12} - p_{23} - m_1^2 + m_3^2)(m_3^2 - m_1^2 + \alpha_{13})}{2p_{13}}, \\
p_{34} &= s_{34}^{(13)},
\end{aligned} \tag{9}$$

and

$$\begin{aligned}
s_{14}^{(ij)} &= \frac{\Delta_{ij} + 2m_i^2 p_{ij} - (p_{ij} + m_i^2 - m_j^2)\alpha_{ij}}{2p_{ij}}, \\
s_{34}^{(ij)} &= \frac{\Delta_{ij} + 2m_j^2 p_{ij} + (p_{ij} + m_j^2 - m_i^2)\alpha_{ij}}{2p_{ij}}, \\
\alpha_{ij} &= \pm\sigma(p_{ij} - m_i^2 + m_j^2)\sqrt{\Delta_{ij}}.
\end{aligned} \tag{10}$$

Again as it was for integral $I_2^{(d)}$ integral $I_3^{(d)}$ with arbitrary arguments can be expressed in terms of integrals with at least one propagator massless. By applying formula (8) to integrals in the right-hand side of this equation several times one can express $I_3^{(d)}$ with arbitrary arguments in terms of integrals with two massless propagators and one line with massive propagator. Also one of kinematical arguments in these integrals will be zero and other arguments will be functions of initial kinematical arguments and masses. This integral can be expressed in terms of Appell hypergeometric function F_1 and Gauss' hypergeometric function ${}_2F_1$ [1].

3. Derivation of FE by using algebraic relations for propagators

Analyzing FE for one-loop Feynman integrals one can observe that integrands are rather similar and differ only by one propagator. In the relation for one-loop propagator type integral integrands are

$$\frac{1}{D_1 D_2}, \ \frac{1}{D_0 D_2}, \ \frac{1}{D_1 D_0},$$

and integrands for the one-loop vertex type integrals are

$$\frac{1}{D_1D_2D_3},\quad \frac{1}{D_0D_2D_3},\frac{1}{D_1D_0D_3},\frac{1}{D_1D_2D_0}$$

where

$$D_0=(k_1-p_0)^2-m_0^2+i\epsilon,\quad D_1=(k_1-p_1)^2-m_1^2+i\epsilon,$$
$$D_2=(k_1-p_2)^2-m_2^2+i\epsilon,\quad D_3=(k_1-p_3)^2-m_3^2+i\epsilon,$$

and $i\epsilon$ is traditional causal prescription. Here k_1 is integration momentum and on scalar products of p_1,p_2,... some restrictions, like $G_n=0$, are imposed. Such restrictions effectively lead to linear dependence of these vectors.

3.1. Algebraic relations for products of propagators

One can raise the question: would it be possible to find algebraic relations for products of propagators and derive FE from such relations? From the explicit FE for one-loop propagator integral we can try to find relation of the form:

$$\frac{1}{D_1D_2}=\frac{x_1}{D_0D_2}+\frac{x_2}{D_1D_0} \tag{11}$$

where we assume that x_1,x_2, p_0, p_1 and p_2 are independent of k_1. Putting all terms over the common denominator we get

$$D_0=x_1D_1+x_2D_2. \tag{12}$$

By differentiating (12) with respect to k_1 we obtain two equations:

$$x_1+x_2=1,\qquad p_0=x_1p_1+x_2p_2. \tag{13}$$

Taking into account these relationships we get from (12) an additional equation:

$$x_1x_2s_{12}+m_0^2-x_1m_1^2-x_2m_2^2=0, \tag{14}$$

where $s_{12}=(p_1-p_2)^2$. From Eqs.(13), (14) one get expressions for x_1,x_2:

$$\begin{aligned} x_1 &= \frac{m_2^2-m_1^2+s_{12}}{2s_{12}}-\frac{\sqrt{\Lambda_2+4s_{12}m_0^2}}{2s_{12}},\\ x_2 &= \frac{m_1^2-m_2^2+s_{12}}{2s_{12}}+\frac{\sqrt{\Lambda_2+4s_{12}m_0^2}}{2s_{12}}, \end{aligned} \tag{15}$$

and

$$\Lambda_2=s_{12}^2+m_1^4+m_2^4-2s_{12}(m_1^2+m_2^2)-2m_1^2m_2^2.$$

Integration of the obtained algebraic relation with respect to k_1 gives the following FE:

$$\begin{aligned} I_2^{(d)}(m_1^2,m_2^2,s_{12}) &= \frac{s_{12}+m_1^2-m_2^2+\lambda}{2s_{12}}\, I_2^{(d)}(m_1^2,m_0^2,s_{13}(m_1^2,m_2^2,m_0^2,s_{12}))\\ &+\frac{s_{12}-m_1^2+m_2^2-\lambda}{2s_{12}}\, I_2^{(d)}(m_2^2,m_0^2,s_{23}(m_1^2,m_2^2,m_0^2,s_{12})), \end{aligned}$$

where

$$s_{13} = \frac{\Lambda_2 + 2s_{12}(m_1^2 + m_0^2)}{2s_{12}} + \frac{m_1^2 - m_2^2 + s_{12}}{2s_{12}}\lambda$$
$$s_{23} = \frac{\Lambda_2 + 2s_{12}(m_2^2 + m_0^2)}{2s_{12}} + \frac{m_1^2 - m_2^2 - s_{12}}{2s_{12}}\lambda.$$
$$\lambda = \sqrt{\Lambda_2 + 4s_{12}m_0^2}$$

Parameter m_0 is arbitrary and can be taken at will. The same equation was obtained from recurrence relations by imposing conditions on Gram determinants.

Similar to the relation with two propagators one can find relation for three propagartors:

$$\frac{1}{D_1D_2D_3} = \frac{x_1}{D_2D_3D_0} + \frac{x_2}{D_1D_3D_0} + \frac{x_3}{D_1D_2D_0}. \tag{16}$$

Here p_0, p_1, p_2 and p_3 are external momenta assumed to be independent on k_1. Momentum k_1 will be integration momentum and we assume also that x_i do not depend on k_1. Multiplying both sides of equation (16) by the product $D_1D_2D_3D_0$ we get

$$D_0 = x_1D_1 + x_2D_2 + x_3D_3. \tag{17}$$

Differentiation of this equation with respect to k_1 gives two equations:

$$x_1 + x_2 + x_3 = 1, \qquad p_0 = x_1p_1 + x_2p_2 + x_3p_3. \tag{18}$$

Taking into account these relationships from Eq. (17) we obtain:

$$x_2x_3p_{23} + x_1x_3p_{13} + x_1x_2p_{12} - x_1m_1^2 - x_2m_2^2 - x_3m_3^2 + m_0^2 = 0. \tag{19}$$

This system has the following solution

$$x_1 = 1 - \alpha - x_2, \qquad x_3 = \alpha, \tag{20}$$

where α is the root of the quadratic equation

$$\alpha^2p_{13} + [m_3^2 - m_1^2 - p_{13} + x_2(p_{13} + p_{12} - p_{23})]\alpha$$
$$+m_1^2 - m_0^2 + (m_2^2 - m_1^2 - p_{12} + p_{12}x_2)x_2 = 0.$$

This root depends on 2 arbitrary parameters - m_0 and x_2. By integrating the obtained relation we get the same FE as it was given before.

As was shown in Ref. [4] one can also derive algebraic relations for products of any number of propagators.

Functional relations for Feynman integrals with integrands being rational functions of scalar products of different momenta to some extent resemble *Abel's addition theorem* [8], [9], [10]. It would be useful to find similarities between Feynman integrals and Abel integrals.

3.2. Some remarks about Abel integrals

In this subsection we will present some facts about Abelian integrals that may be useful in investigation of functional equations for Feynman integrals. Abelian integral is an integral in the complex plane of the form

$$\int_{z_0}^{z} R(x, w)dx,$$

where $R(x, w)$ is an arbitrary *rational* function of the two variables x and w. These variables are related by the equation

$$F(x, w) = 0,$$

where $F(x, w)$ is an irreducible polynomial in w,

$$F(x, w) \equiv \phi_n(x)w^n + ... + \phi_1(x)w + \phi_0(x),$$

whose coefficients $\phi_j(x)$, $j = 0, 1, ...n$ are rational functions of x. Abelian integrals are natural generalizations of elliptic integrals, which arise when

$$F(x, w) = w^2 - P(x),$$

where $P(x)$ is a polynomial of degree 3 and 4. If degree of the polynomial is greater than 4 then we have *hyperelliptic integral.* For Abel integrals an important theorem was proven.
Let C and C' be plane curves given by the equations

$$\begin{aligned} C: &\quad F(x, y) = 0, \\ C': &\quad \phi(x, y) = 0. \end{aligned}$$

These curves have n points of intersections $(x_1, y_1), \ldots$ (x_n, y_n), where n is the product of degrees of C and C'. Let $R(x, y)$ be a rational function of x and y where y is defined as a function of x by the relation $F(x, y) = 0$. Consider the sum

$$I = \sum_{i=1}^{n} \int_{x_0, y_0}^{x_i, y_i} R(x, y) dx. \tag{21}$$

Integrals being taken from a fixed point to the n points of intersections of C and C'. If some of the coefficients $a_1, a_2, \ldots, a_k$ of $\phi(x, y)$ are regarded as continuous variables, the points (x_i, y_i) will vary continuously and hence I will be a function, whose form is to be determined, of the variable coefficients $a_1, a_2, \ldots, a_k$.
Abel's theorem:
The partial derivatives of the sum I, with respect to any of the coefficients of the variable curve $\phi(x, y) = 0$, is a *rational* function of the coefficients and hence I is equal to a *rational* function of the coefficients of $\phi(x, y) = 0$, plus a finite number of logarithms or arc tangents of such rational functions. What is very important - integrals themselves can be rather complicated transcendental functions but their sum can be simple.
Example: Elliptic integral of the second type [9]:

$$E(k, x) = \int_0^x \frac{(1 - k^2 x)dx}{\sqrt{x(1 - x)(1 - k^2 x)}}.$$

We take as C and C'

$$C: \quad y^2 = x(1 - x)(1 - k^2 x), \qquad\qquad C': \quad y = ax + b. \tag{22}$$

The elimination of y between two equations will give us as the abscissae x_1, x_2, x_3 of the points of intersection the three roots of the equation:

$$\phi(x) = k^2 x^3 - (1 + k^2 + a^2)x^2 + (1 - 2ab)x - b^2 = 0. \tag{23}$$

The corresponding sum will be

$$I(a,b) = \int_0^{x_1} R(x,y)dx + \int_0^{x_2} R(x,y)dx + \int_{1/k^2}^{x_3} R(x,y)dx,$$

where

$$R(x,y) = \frac{1-k^2x}{y}. \tag{24}$$

Abel's theorem gives addition formula:

$$\int_0^{x_1} R(x)dx + \int_0^{x_2} R(x)dx + \int_{1/k^2}^{x_3} R(x)dx = -2a + \kappa, \tag{25}$$

where κ is an arbitrary constant.

One can see that FE for Feynman integrals and the relationship (25) have some common features. Arguments of Feynman integrals and arguments of the functions in the left-hand side of Eq. (25) are determined from algebraic equations and integrands in both cases are rational functions of some algebraic functions.

One can find FE for Feynman integrals following closely derivation of relationships for usual algebraic integrals. Deriving relations for propagators we used orthogonality condition $G_n = 0$. In fact it is not needed to assume such a relation. For example, to fix parameters in algebraic relations for products of two propagators

$$R_2(k_1, p_1, p_2, m_1^2, m_2^2, m_0^2) = \frac{1}{D_1D_2} - \frac{x_1}{D_2D_0} - \frac{x_2}{D_1D_0} = 0, \tag{26}$$

instead of $G_n = 0$ we can impose conditions

$$\frac{\partial x_1}{\partial k_{1\mu}} = \frac{\partial x_2}{\partial k_{1\mu}} = 0, \tag{27}$$

having in mind that only D_j factors will depend on momentum k_1. Multiplying both sides of Eq. (26) by $D_0D_1D_2$ we get

$$\begin{aligned} &D_0 - x_1D_1 - x_2D_2 \\ &= (1-x_1-x_2)k_1^2 + 2x_1k_1p_1 + 2x_2k_1p_2 \\ &+x_1m_1^2 + x_2m_2^2 - x_1p_1^2 - x_2p_2^2 - 2k_1p_0 - m_0^2 + p_0^2 = 0. \end{aligned} \tag{28}$$

Differentiating this relation with respect to $k_{1\mu}$, contracting with k_1, p_1, p_2, p_0 and taking into account (27) gives several equations:

$$\begin{aligned} &-2x_1(k_1^2 - k_1p_1) - 2x_2(k_1^2 - k_1p_2) + 2k_1^2 - 2k_1p_0 = 0, \\ &2(1-x_1-x_2)k_1p_1 + 2x_1p_1^2 + 2x_2p_1p_2 - 2p_1p_0 = 0, \\ &2(1-x_1-x_2)k_1p_2 + 2p_1p_2x_1 + 2x_2p_2^2 - 2p_2p_0 = 0, \\ &2(1-x_1-x_2)k_1p_0 + 2x_1p_1p_0 + 2x_2p_2p_0 - 2p_0^2 = 0. \end{aligned} \tag{29}$$

They can be used to express k_1p_0, p_1p_0, p_2p_0, x_1, x_2 in terms of k_1^2, k_1p_1, k_1p_2, p_1^2, p_1p_2, p_2^2 considered to be independent variables. For example, we get:

$$k_1p_0 = x_1k_1p_1 - x_1k_1p_2 + k_1p_2 + \frac{x_1}{2}(m_1^2 - m_2^2 - p_1^2 + p_2^2) + \frac{1}{2}(m_2^2 - p_2^2 - m_0^2 + p_0^2), \tag{30}$$

and similar expressions for other scalar products of p_0. Solution for x_1, x_2 is the same as in Eq. (15) and as a result we obtained the same relation between products of two propagators.

Solution of the above system of equations is rather similar to finding intersections of two plane curves considered in Abel's theorem. Unfortunately extension of Abel's theorem for integrals of algebraic functions of several variables is not an easy task. In the case of functions of two variables some results were obtained long time ago in Refs. [11], [12].

Similar to usual algebraic integrals of one variable we can construct various integrands out of our different relationships for products of propagator. These integrands will be rational functions in independent variables. Integrations should be done over d dimensional space. Rational function must resemble integrands for Feynman integrands.

For example, multiplying $R_2(k_2, p_2, p_4, m_2^2, m_4^2, \widetilde{m}_0^2)$ by

$$\frac{1}{[(k_1-p_1)^2-m_1^2]^{\nu_1}[(k_1-p_3)^2-m_3^2]^{\nu_3}[(k_1-k_2)^2-m_5^2]^{\nu_5}}, \tag{31}$$

and integrating this product with respect to k_1, k_2 leads to the following FE:

$$\int \frac{d^dk_1 d^dk_2 \quad R_2(k_2, p_2, p_4, m_2^2, m_4^2, \widetilde{m}_0^2)}{[(k_1-p_1)^2-m_1^2]^{\nu_1}[(k_1-p_3)^2-m_3^2]^{\nu_3}[(k_1-k_2)^2-m_5^2]^{\nu_5}} = 0, \tag{32}$$

Integrals in this equation correspond to the diagram given in Fig.1.

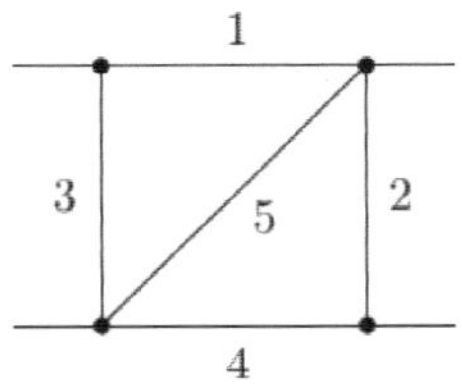

Fig.1. FE for the two-loop box integral.

By integrating product of the relationship

$$R_3(k_1, p_1, p_2, p_3, m_1^2, m_2^2, m_3^2, m_0^2) = \frac{1}{D_1D_2D_3} - \frac{x_1}{D_2D_3D_0} - \frac{x_2}{D_1D_3D_0} - \frac{x_3}{D_1D_2D_0} = 0,$$

and one loop-propagator integral

$$\int\int \frac{d^dk_1 d^dk_2}{[(k_1-p_1)^2-m_1^2]^{\nu_1}[(k_1-k_2)^2-m_5^2]^{\nu_5}} R_3(k_2, p_2, p_3, p_4, m_2^2, m_3^2, m_4^2, m_0^2) = 0,$$

we obtain the FE for the integral corresponding to the following diagram with arbitrary ν_1, ν_5, and arbitrary momenta and masses

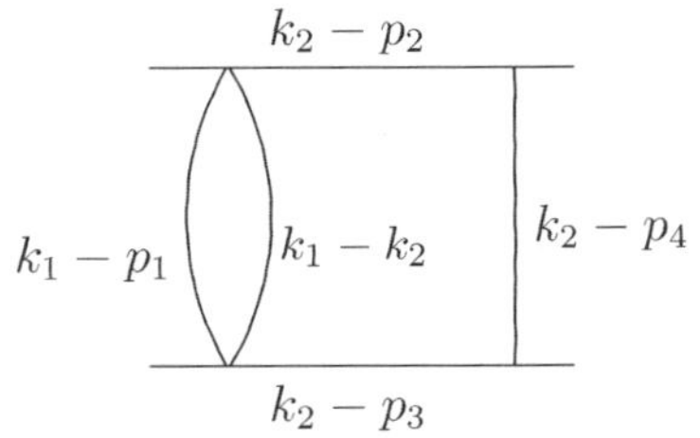

Fig.2. FE for the two-loop integral with propagator insertion.

By integrating different products of R_2, R_3 and products of different propagators with respect to selected momenta one can easily derive FE for various multiloop Feynman integrals.

IOP Conf. Series: Journal of Physics: Conf. Series **920** (2017) 012004 doi:10.1088/1742-6596/920/1/012004

4. Concluding remarks

We described two different methods for finding FE for Feynman integrals with any number of loops and external legs. FE reduce integrals with complicated kinematics to simpler integrals. FE can be used for analytic continuation of Feynman integrals without knowing explicit analytic result. At the present time application of these methods for some two- and three- loop integrals is in progress. Systematic investigation of FE for Feynman integrals based on algebraic geometry and group theory is needed. Some improvements of these methods can be done by exploiting known methods for algebraic integrals. The methods can be extended for finding functional equations among hypergeometric as well as holonomic functions.

5. Acknowledgment

I would like to thank organizers of the CPP2016 for the invitation and great workshop. My contribution to the proceedings I dedicate to the memory of Shimizu - Sensei. I am also thankful to referee for useful suggestions taken into account in Section 3 of this paper.

6. References

[1] Tarasov O V 2008 New relationships between Feynman integrals. *Phys.Lett.*, B670:67–72
[2] Tarasov O V 2011 Functional equations for Feynman integrals. *Phys.Part.Nucl.Lett.*, 8:419–427
[3] Kniehl B A and Tarasov O V 2009 Functional equations for one-loop master integrals for heavy-quark production and Bhabha scattering. *Nucl.Phys.*, B820:178–192
[4] Tarasov O V 2015 Derivation of Functional Equations for Feynman Integrals from Algebraic Relations. *Preprint* hep-ph/1512.09024
[5] Tarasov O V 1996 Connection between Feynman integrals having different values of the space-time dimension. *Phys.Rev.*, D54:6479–6490.
[6] Bollini C G and Giambiagi J J 1972 Lowest order divergent graphs in nu-dimensional space. *Phys. Lett.*, B40:566–568
[7] Boos E E and Davydychev A I 1991 A Method of evaluating massive Feynman integrals. *Theor.Math.Phys.*, 89:1052–1063
[8] Markushevich A I 1992 *Introduction to the Classical Theory of Abelian Functions.* Translations of mathematical monographs. American Mathematical Society
[9] Barnum H H 1910 Abel's theorem and the addition formulae for elliptic integrals. *Annals of Mathematics*, 11(3):103–114
[10] Vilenkin N J and Klimyk A U 1992 *Representation of Lie Groups and Special Functions: Volume 3: Classical and Quantum Groups and Special Functions.* Mathematics and its Applications. Springer Netherlands
[11] Picard E and Simart G 1897,1900, 1904,1906 *Theorie des fonctions algebriques de deux variables independantes*, volume 1,2.
[12] Forsyth A R 1882 On Abel's theorem and abelian functions. *Proceedings of the Royal Society of London*, 34:288–291

IOP Conf. Series: Journal of Physics: Conf. Series **920** (2017) 012005 doi:10.1088/1742-6596/920/1/012005

Numerical integration and extrapolation for finite and UV-divergent 3-loop Feynman integrals

E de Doncker[1], F Yuasa[2], K Kato[3] and T Ishikawa[2]

[1] Department of Computer Science, Western Michigan University, Kalamazoo MI 49008, U. S. A.
[2] High Energy Accelerator Research Organization (KEK), Oho 1-1, Tsukuba, Ibaraki, 305-0801, Japan
[3] Kogakuin University, Shinjuku, Tokyo 163-8677, Japan

E-mail:
elise.dedoncker@wmich.edu, fukuko.yuasa@kek.jp, kato@cc.kogakuin.ac.jp, tadashi.ishikawa@kek.jp

Abstract. Automatic numerical integration attempts to alleviate the detailed application of solution methods to the problem. We use automatic multivariate integration with PARINT combined with numerical extrapolation for classes of 3-loop UV-divergent self-energy diagrams with massive internal lines, and finite or UV-divergent massless vertex diagrams. The extrapolation is performed with respect to the dimensional regularization parameter. It is shown that the PARINT software handles the integrations accurately and efficiently.

1. Introduction

The radiative correction (RC) is highly important to study the physics of high-precision experiments at the current and future colliders. The development of methods to compute the integral in a multi-loop Feynman diagram is indispensable for the higher-order RC.

For a Feynman diagram with L loops and N internal lines, we consider the integral

$$\begin{aligned} \mathcal{I} &= \Gamma(N-nL/2)(-1)^N \int_0^1 \prod_{r=1}^{N} dx_r\, \delta(1-\sum x_r)\, \frac{C^{N-n(L+1)/2}}{(D-i\varrho C)^{N-nL/2}} \\ &= \Gamma(N-nL/2)(-1)^N \int_{\mathcal{S}_{N-1}} \frac{C^{N-n(L+1)/2}}{(D-i\varrho C)^{N-nL/2}}\, d\mathbf{x} \qquad (1) \end{aligned}$$

where C and D are polynomials determined by the topology of the corresponding diagram and physical parameters, $n = 4 - 2\varepsilon$ to account for UV singularity (where ε is the dimensional regularization parameter), and

$$\mathcal{S}_d = \{\mathbf{x} \in \mathbb{R}^d \mid \sum_{r=1}^{d} x_r \le 1, x_r \ge 0 \text{ for } 1 \le r \le d\}$$

is the d-dimensional unit simplex. In the absence of singularities in the interior of the integration domain we set $\varrho = 0$. Feynman loop integrals defined in Eq (1) can be expanded in a Laurent series with respect to the dimensional regularization parameter ε, of the form $\mathcal{I} \sim \sum_{k \ge K} C_k\, \varepsilon^k$, for an integer K. In this paper we obtain the leading coefficients C_k of the expansion numerically using automatic integration and linear extrapolation.

Automatic integration implements a *black-box* approach to obtain an approximation Qf to an integral

$$If = \int_{\mathcal{D}} f(\mathbf{x})\, \mathbf{dx} \tag{2}$$

and an absolute error estimate E_a, in order to satisfy a specified accuracy requirement for the actual error, of the form

$$|\, Qf - If \,| \;\le\; E_a \;\le\; \max\{\, t_a \,,\, t_r\, |If| \,\} \tag{3}$$

for a given integrand function f, region $\mathcal{D}$ and (absolute/relative) error tolerances t_a and t_r. We denote the corresponding relative error estimate by E_r, satisfying $E_a = E_r\, |Qf|$. The type of accuracy requirement given by Eq (3) is based on [1] and used extensively in QUADPACK [2]. The integral of Eq (1) will be transformed so the domain $\mathcal{D}$ is the $(N-1)$-dimensional unit cube.

For the numerical integrations in this paper we apply the global adaptive code in the PARINT package (see, e.g., [3] for a description of PARINT), which implements a region partitioning strategy layered over MPI [4] and generally distributed over a network of computational nodes. The global adaptive strategy sums the total absolute error estimate E_a contributed over the region partition. On each subregion we use the integral and error estimates given for the multivariate cube in [5]. The integration and error estimation over the subregions is performed by the worker processes, while giving preference to the highest-error regions for subdivision. The controller process evaluates the overall termination criteria and (by default) also participates in the local integrations. For termination, the second inequality of Eq (3) is approximated by

$$E_a \;\le\; \max\{\, t_a \,,\, t_r\, |Qf| \,\} \tag{4}$$

where Qf is the total integral approximation. Apart from Eq (4), termination is also triggered when the user-specified maximum number of integrand evaluations is reached (whichever comes first). The latter refers to the total number of times the integrand function sub-program is called, and is proportional to the overall number of region subdivisions (also totaled by the controller over all worker processes). The combination of the termination criteria and the importance-driven subdivisions tailored to the specific difficulties in the integrand function lead to a fully automatic and machine-learned integration procedure.

With respect to expanding Eq (1) let us first consider a general expansion

$$S(\varepsilon) \sim C_K\, \varphi_K(\varepsilon) + C_{K+1}\, \varphi_{K+1}(\varepsilon) + C_{K+2}\, \varphi_{K+2}(\varepsilon) + \ldots \tag{5}$$

where we assume the φ_k functions are known (for example, $\varphi_k(\varepsilon) = \varepsilon^k$). To approximate the leading coefficients $C_k,\ k = K, K+1, \ldots$, we apply a linear extrapolation, yielding the coefficients by solving a linear system based on the underlying expansion. This is achieved by generating a sequence of $S(\varepsilon_\ell)$ values such that

$$S(\varepsilon_\ell) = C_K\, \varphi_K(\varepsilon_\ell) + C_{K+1}\, \varphi_{K+1}(\varepsilon_\ell) + \ldots C_{K+\nu}\, \varphi_{K+\nu}(\varepsilon_\ell), \quad \ell = k, \ldots, \nu + k \tag{6}$$

$(\nu, k \ge 0)$ and solving the linear system of order $(\nu+1) \times (\nu+1)$. For error control this can be done for increasing values of ν and decreasing $\varepsilon = \varepsilon_\ell$. For example, a geometric sequence $\varepsilon_\ell = b^{-\ell}, b > 1$ can be used, or a Bulirsch [6] type sequence of the form $\varepsilon_\ell = 1/b_\ell,\ b_\ell = 2, 3, 4, 6, 8, 12, 16, 24, \ldots$.

Non-linear extrapolation with the ϵ-algorithm [7, 8, 9] can be applied (for unknown $\varphi_k(\varepsilon)$) under more general conditions with geometric sequences of ε_ℓ. In previous work we have made ample use of the ϵ-algorithm (see for instance, [10, 11, 12, 13]) to treat loop integrals in the physical region, with integrand singularities caused by vanishing denominators in the interior of the integration domain. In this paper we employ linear extrapolation to obtain the coefficients of the Laurent series.

The method is demonstrated in Section 2 for two self-energy diagrams, where integrand transformations are further introduced to diminish singularities at the boundaries of the domain. Section 3 presents results for 3-loop vertex diagrams, and Section 4 applies an extrapolation to a 2-fold Mellin-Barnes integral representation.

2. 3-loop self-energy diagrams

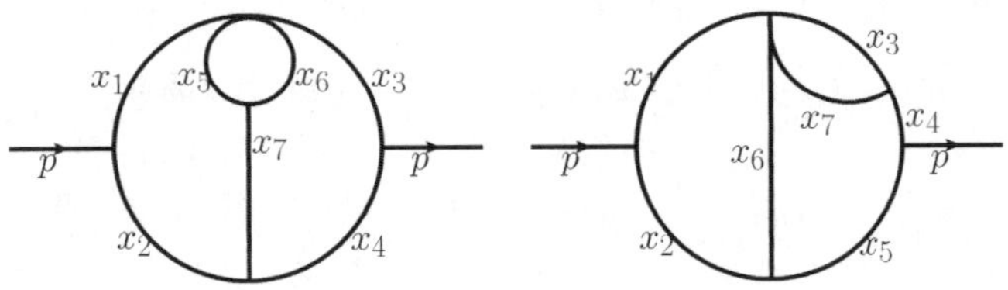

(a) (b)

Figure 1. 3-loop UV-divergent self-energy diagrams with massive internal lines ($m_i = 1$ and $s = p^2 = 1$), cf., Laporta [14]: (a) Laporta Fig 2(o) ($N = 7$, *ladybug* diagram), (b) Laporta Fig 2(p) ($N = 7$, *cracked Magdeburg* diagram)

Corresponding to the *ladybug* diagram of Fig 1(a) and the *cracked Magdeburg* diagram of Fig 1(b) with $L = 3$ and $N = 7$, we consider the 3-loop self-energy integrals

$$I^{S3}(\varepsilon) = \Gamma(1+3\varepsilon)\int_{S_6} \frac{C^{-1+4\varepsilon}}{D^{1+3\varepsilon}}\, d\mathbf{x} \tag{7}$$

The C and D functions derived from the diagrams are the determinants given in Table (1), where we denote $x_{j_1 j_2 \dots j_k} = x_{j_1} + x_{j_2} + \dots + x_{j_k}$. The corresponding polynomials are given as $(C_a^{S3},\ D_a^{S3})$ in Eq (A.1), and $(C_b^{S3},\ D_b^{S3})$ in Eq (A.2) of Appendix A in this paper.

Table 1. C and D functions of self-energy integrals (7)

I_a^{S3}	I_b^{S3}
$C = \begin{vmatrix} x_{1257} & -x_7 & -x_5 \\ -x_7 & x_{3467} & -x_6 \\ -x_5 & -x_6 & x_{56} \end{vmatrix}$	$C = \begin{vmatrix} x_{126} & -x_6 & 0 \\ -x_6 & x_{4567} & -x_7 \\ 0 & -x_7 & x_{37} \end{vmatrix}$
$D = -\begin{vmatrix} x_{1257} & -x_7 & -x_5 & px_1 \\ -x_7 & x_{3467} & -x_6 & px_3 \\ -x_5 & -x_6 & x_{56} & 0 \\ px_1 & px_3 & 0 & p^2x_{13} - \sum_{i=1}^{7} m_i^2 x_i \end{vmatrix}$	$D = -\begin{vmatrix} x_{126} & -x_6 & 0 & px_1 \\ -x_6 & x_{4567} & -x_7 & px_4 \\ 0 & x_7 & x_{37} & px_3 \\ px_1 & px_4 & px_3 & p^2x_{134} - \sum_{i=1}^{7} m_i^2 x_i \end{vmatrix}$

The integrals (7) satisfy an asymptotic expansion $\sum_{k\geq -1} C_k\, \varepsilon^k$ of the form (5), corresponding to [14] (2o) and (2p):

$$\Gamma^{-3}(1+\varepsilon)\, I_a^{S3}(\varepsilon) \sim 0.92363182652\,\varepsilon^{-1} - 2.423491634 + 8.38134971\,\varepsilon - 26.9936212\,\varepsilon^2 \dots \tag{8}$$

$$\Gamma^{-3}(1+\varepsilon)\, I_b^{S3}(\varepsilon) \sim 0.92363182652\,\varepsilon^{-1} - 2.116169719 + 6.92954468\,\varepsilon - 21.5032784\,\varepsilon^2 \dots \tag{9}$$

Note that numerical results for these integrals were given in [15] using a product trapezoidal rule with double exponential transformation (DE) on Intel(R) Xeon(R) E5-2687W v3 3.10 GHz in IEEE 754-2008 binary128, which required extensive computation times. However the executions with PARINT in long double precision reported later in this section take far less than the DE time (while reaching near comparable accuracy).

For the numerical approximation of the integrals over $\mathcal{S}_{N-1}$, a variable transformation is performed that maps the simplex domain to the unit cube and at the same time achieves a smoothening of the integrand boundary singularities. For I_a^{S3} the variables are transformed as

$$\begin{aligned} x_1 &= y_{1m}\, y_2\, y_4\, y_5, \quad x_2 = y_{1m}\, y_2\, y_4\, y_{5m}, \\ x_3 &= y_{1m}\, y_2\, y_{4m}\, y_6, \quad x_4 = y_{1m}\, y_2\, y_{4m}\, y_{6m}, \\ x_5 &= y_1\, y_3, \quad x_6 = y_1\, y_{3m}, \\ x_7 &= y_{1m}\, y_{2m} \end{aligned} \tag{10}$$

with $y_{im} = 1 - y_i$ and Jacobian $y_1\, y_{1m}^4\, y_2^3\, y_4\, y_{4m}$. The transformation applied for I_b^{S3} is

$$\begin{aligned} x_1 &= y_1\, y_2\, y_4, \quad x_2 = y_1\, y_2\, y_{4m}, \\ x_3 &= y_{1m}\, y_3\, y_6, \quad x_4 = y_1\, y_{2m}\, y_5, \\ x_5 &= y_1\, y_{2m}\, y_{5m}, \quad x_6 = y_{1m}\, y_{3m}, \\ x_7 &= y_{1m}\, y_3\, y_{6m} \end{aligned} \tag{11}$$

with $y_{im} = 1 - y_i$ and Jacobian $y_1^3\, y_{1m}^2\, y_{2m}\, y_2\, y_3$. While the above two variable transformations are interchangeable for I_a^{S3} and I_b^{S3}, the transformation (10) yields better results for I_a^{S3}, and (11) for I_b^{S3}.

Numerical integration results were obtained with PARINT on the *thor* cluster at WMU, using 64 processes on four 16-core nodes (each with dual Intel Xeon E5-2670, 8C, 2.6 GHz processors, and 128 GB memory). We employ the adaptive method (over the 6D cube) with the multivariate cubature rule of polynomial degree 9 in [5] for the local integration over the subregions. The convergence of linear extrapolation applied to the PARINT sequence as $\varepsilon = \varepsilon_\ell \to 0$ is shown in Table 2 and Table 3 for (8) and (9), respectively. The computation of each integral as a function of ε, corresponding to each row, is thus performed automatically according to the global adaptive strategy. The tables also list the obtained estimated relative error $E_r = E_a/|Qf|$. Table 2 further shows the user-specified maximum number of function evalutions MAX given in billions (B) for each integration. Setting a strict relative error tolerance ($t_r = 10^{-13}$) allows terminating based on MAX, which is the case for the computations in Table 2 and part of Table 3. The latter gives the actual number of function evaluations (#EV) performed for the integration. Since for the UV self-energy problems at hand, the integration becomes harder as ε_ℓ decreases, the relative error estimates E_r tend to increase for successive ℓ.

Table 2. Results 3-loop UV self-energy integral $I_a^{S3}(\varepsilon)\,\Gamma^{-3}(1+\varepsilon)$ (on 4 nodes with 16 procs/node of *thor* cluster, *long double* precision); integration err. tolerances $t_r = 10^{-13}$, $t_a = 0$; $E_r = E_a/|Qf|$ (= integration estim. rel. error); #EV[B] = # integrand evals. in billons; T[s] = Time (elapsed user time) in seconds; $\varepsilon = \varepsilon_\ell = 1.15^{-\ell}$

	INTEGRAL Fig 1 (a)			EXTRAPOLATION			
ℓ	E_r	MAX[B]	T[s]	RESULT C_{-1}	RESULT C_0	RESULT C_1	RESULT C_2
17	2.3e-12	100	682.3				
18	4.2e-12	100	682.8	0.88532153209	-1.412495535		
19	6.0e-12	100	683.8	0.91647848504	-2.133365157	4.1493435	
20	5.2e-12	150	1027.6	0.92246162843	-2.356946186	6.9162602	-11.33999
21	2.2e-12	350	2415.8	0.92346458479	-2.410840073	7.9934802	-20.83203
22	3.6e-12	350	2419.5	0.92361089965	-2.421456174	8.2986081	-25.17454
23	5.7e-12	350	2425.8	0.92362953012	-2.423211188	8.3667190	-26.56843
24	8.7e-12	350	2431.6	0.92363160796	-2.423458644	8.3791875	-26.91297
25	1.3e-11	350	2437.1	0.92363180412	-2.423487620	8.3810333	-26.97920
26	1.7e-11	350	2445.7	0.92363182397	-2.423491206	8.3813166	-26.99204
			Exact (8):	0.92363182652	-2.423491634	8.3813497	-26.99362

Table 3. Results 3-loop UV self-energy integral $I_b^{S3}(\varepsilon)\,\Gamma^{-3}(1+\varepsilon)$ (on 4 nodes with 16 procs/node of *thor* cluster, *long double* precision); integration err. tolerances $t_r = 10^{-13}$, $t_a = 0$; $E_r = E_a/|Qf|$ (= integration estim. rel. error); #Ev[B] = # integrand evals. in billons; T[s] = Time (elapsed user time) in seconds; $\varepsilon = \varepsilon_\ell = 1.3^{-\ell}$

	Integral Fig 1 (b)				Extrapolation		
ℓ	E_r	#Ev[B]	T[s]	Result C_{-1}	Result C_0	Result C_1	Result C_2
5	7.1e-14	121	915.1				
6	5.7e-14	95	668.8	0.79556987526	-0.624745066		
7	6.4e-14	110	760.4	0.88129897281	-1.356849184	1.5364029	
8	3.9e-13	200	1444.9	0.91188370051	-1.809949406	3.7234312	-3.43942
9	2.2e-13	275	2104.7	0.92097424134	-2.018776382	5.4720364	-9.76427
10	2.4e-12	275	2018.5	0.92314713425	-2.091734298	6.4193304	-15.70840
11	7.0e-12	275	1978.3	0.92356124561	-2.111347543	6.7915499	-19.33007
12	1.3e-11	325	2288.4	0.92362368266	-2.115423673	6.9006678	-20.88190
13	3.2e-11	325	2309.7	0.92363108618	-2.116079491	6.9248672	-21.15424
14	6.7e-11	325	2331.9	0.92363177196	-2.116161010	6.9289487	-21.36736
15	1.4e-10	350	2534.4	0.92363182585	-2.116169538	6.9295216	-21.50172
			Exact (9):	0.92363182652	-2.116169718	6.9295447	-21.50328

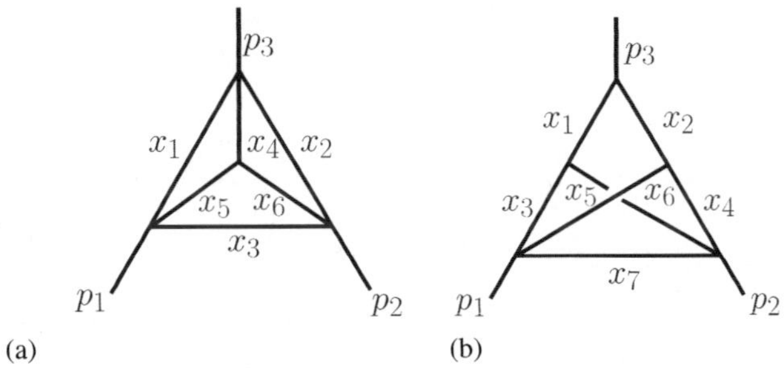

Figure 2. 3-loop vertex diagrams with massless internal lines, cf., Heinrich et al. [16] (a) Diagram $A_{6,2}$ ($N = 6$), (b) Diagram $A_{7,5}$ ($N = 7$). In [16] the incoming momentum is $p_3 = p_1 + p_2$ and outgoing lines are on-shell and massless, $p_1^2 = p_2^2 = 0$.

3. 3-loop vertex diagrams

For the diagrams in Fig 2(a) and (b) with $N = 6$ and $N = 7$, respectively, we denote

$$I_a^{V3}(\varepsilon) = \Gamma(3\varepsilon) \int_{\mathcal{S}_5} \frac{C^{-2+4\varepsilon}}{D^{3\varepsilon}}\, d\mathbf{x} \tag{12}$$

$$I_b^{V3}(\varepsilon) = -\Gamma(1+3\varepsilon) \int_{\mathcal{S}_6} \frac{C^{-1+4\varepsilon}}{D^{1+3\varepsilon}}\, d\mathbf{x} \tag{13}$$

The C and D functions in the integrands are listed as determinants in Table 4 for Eq (12) and in Table 5 for Eq (13). Here $p_1^2 = p_2^2 = 0$, so $-q^2 = -p_3^2 = -(p_1+p_2)^2 = -2p_1p_2$ (which is written as a separate factor outside of the computed integral). Expanded versions of C and D are given as $(C_a^{V3},\ D_a^{V3})$ in Eq (A.3), and $(C_b^{V3},\ D_b^{V3})$ in Eq (A.4) of Appendix A.

Table 4. C and D functions of vertex integral (12), $p_1^2 = p_2^2 = 0$

$$I_a^{V3}$$

$$C = \begin{vmatrix} x_{145} & -x_4 & -x_5 \\ -x_4 & x_{246} & -x_6 \\ -x_5 & -x_6 & x_{356} \end{vmatrix}$$

$$D = -\begin{vmatrix} x_{145} & -x_4 & -x_5 & p_1x_1 \\ -x_4 & x_{246} & -x_6 & -p_2x_2 \\ -x_5 & -x_6 & x_{356} & 0 \\ p_1x_1 & -p_2x_2 & 0 & p_1^2x_1 + p_2^2x_2 \end{vmatrix}$$

Table 5. C and D functions of vertex integral (13), $p_1^2 = p_2^2 = 0$

$$I_b^{V3}$$

$$C = \begin{vmatrix} x_{1236} & x_{12} & x_3 \\ x_{12} & x_{1245} & -x_5 \\ x_3 & -x_5 & x_{357} \end{vmatrix}$$

$$D = -\begin{vmatrix} x_{1236} & x_{12} & x_3 & p_1x_{13} - p_2x_2 \\ x_{12} & x_{1245} & -x_5 & p_1x_2 - p_2x_{24} \\ x_3 & -x_5 & x_{357} & p_1x_3 \\ p_1x_{13} - p_2x_2 & p_1x_1 - p_2x_{24} & p_1x_3 & p_1^2x_{13} + p_2^2x_{24} \end{vmatrix}$$

Table 6. Results 3-loop UV vertex integral of Eq (14) by PARINT (on 4 nodes with 16 procs/node of *thor* cluster, *long double* precision); integration err. tolerances $t_r = 10^{-13}$, $t_a = 0$; MAX = 30B (= max. # integrand evals.); $E_r = E_a/|Qf|$ (= integration estim. rel. error); T[s] = Time (elapsed user time) in seconds; $\varepsilon = \varepsilon_\ell = 1.2^{-\ell}$

	INTEGRAL Fig 2(a)					
ℓ	E_r	T[s]	RES. C_{-1}	RES. C_0	RES. C_1	RES. C_2
20	6.9e-09	171.6				
21	4.7e-09	171.6	-2.2674024847	-36.22875015		
22	3.8e-09	171.2	-2.4184908276	-23.48554838	-26.647842	
23	4.2e-09	173.2	-2.4030392631	-25.64179665	-16.727996	-150.451
24	4.0e-09	174.5	2.4041730407	-25.40846949	-185.040255	-911.900
25	3.5e-09	175.9	-2.4041116670	-25.42597896	-183.074773	-1020.406
26	3.8e-09	177.3	-2.4041138550	-25.42514605	-183.204386	-1009.854
	Exact (14):		-2.4041138063	-25.42515557	-183.204184	-1009.791

Asymptotic expansions are derived analytically for $I_a^{V3}(\varepsilon)$ and $I_b^{V3}(\varepsilon)$ in [16]:

$$\begin{aligned} -\Gamma^3(1-\varepsilon)\, I_a^{V3}(\varepsilon) \;\sim\; & -2\zeta_3\,\varepsilon^{-1} - (18\zeta_3 + \frac{7\pi^4}{180}) + (-122\zeta_3 - \frac{7\pi^4}{20} + \frac{2\pi^2}{3}\zeta_3 - 10\zeta_5)\,\varepsilon \\ & + (-738\zeta_3 - \frac{427\pi^4}{180} + 6\pi^2\zeta_3 - 90\zeta_5 + \frac{163\pi^6}{7560} + 76\zeta_3^2)\,\varepsilon^2 + \mathcal{O}(\varepsilon^3) \\ \approx\; & -2.4041138063\,\varepsilon^{-1} - 25.42515557 - 183.204184\,\varepsilon - 1009.790681\,\varepsilon^2 + \mathcal{O}(\varepsilon^3) \end{aligned} \tag{14}$$

Table 7. Results 3-loop UV vertex integral of Eq (15) by PARINT (on 4 nodes with 16 procs/node of *thor* cluster, *long double* precision); integration err. tolerances $t_r = 10^{-13}$, $t_a = 0$; MAX = 100B (= max. # integrand evals.); $E_r = E_a/|Qf|$ (= integration estim. rel. error); T[s] = Time (elapsed user time) in seconds; $\varepsilon = \varepsilon_\ell = 1.2^{-\ell}$

	INTEGRAL Fig 2(b)			
ℓ	E_r	T[s]	RES. C_0	RES. C_1
25	7.8e-07	667.5		
26	6.8e-07	667.4	33.8889738	338.4560
27	6.0e-07	667.5	34.1049542	293.1279
28	5.6e-07	667.4	34.0967447	295.9785
29	5.3e-07	667.4	34.0969222	295.8877
	Exact (15):		34.0969298	295.8700

and

$$
\begin{aligned}
-\Gamma^3(1-\varepsilon)\, I_b^{V3}(\varepsilon) &\sim 2\pi^2\zeta_3 + 10\zeta_5 + (12\pi^2\zeta_3 + 60\zeta_5 + \frac{11\pi^6}{162} + 18\zeta_3^2)\,\varepsilon + \mathcal{O}(\varepsilon^2) \\
&\approx 34.0969298 + 295.8700\,\varepsilon + \mathcal{O}(\varepsilon^2)
\end{aligned} \tag{15}
$$

where the $(-)$ sign on the left accounts for a difference in the sign with the definitions of [16].

The integral of Eq (12) is approximated by PARINT and Table 6 lists linear extrapolation results for the sequence (14) as $\varepsilon = \varepsilon_\ell \to 0$. Table 7 shows an extrapolation of a sequence of PARINT results for Eq (15). The integrations are performed using the adaptive method with the rules of polynomial degree 9 in PARINT after mapping the $d = (N-1)$-dimensional simplex to the cube with the transformation

$$
(x_1, \ldots, x_d) \to (y_1, (1-x_1)y_2, \ldots, (1-x_1-\ldots-x_{d-1})y_d) \tag{16}
$$

where the coordinates y_k satisfy $0 \le y_k \le 1$.

4. Mellin-Barnes representation

Using a 2-fold Mellin-Barnes representation from Heinrich et al. [16], we consider the integral

$$
\begin{aligned}
-\Gamma^3(1-\varepsilon) I_a^{V3}(\varepsilon) &= -\frac{\Gamma^3(1-\varepsilon)\,\Gamma(3\varepsilon)\,\Gamma^2(1-3\varepsilon)}{\Gamma(1-2\varepsilon)\,\Gamma(2-4\varepsilon)} \int_{c_1-i\infty}^{c_1+i\infty} \frac{dw_1}{2\pi i} \int_{c_2-i\infty}^{c_2+i\infty} \frac{dw_2}{2\pi i} \\
&\times \frac{\Gamma(-1+3\varepsilon-w_1)\,\Gamma(-1+2\varepsilon-w_1)\,\Gamma(2-4\varepsilon+w_1)\Gamma(-w_2)\,\Gamma(w_2-w_1)}{\Gamma(3\varepsilon-w_1)\,\Gamma(2-4\varepsilon+w_2)\,\Gamma(2-4\varepsilon+w_1-w_2)} \\
&\times \Gamma(1-\varepsilon+w_2)\,\Gamma(1-\varepsilon+w_1-w_2)\,\Gamma(1-2\varepsilon+w_2)\,\Gamma(1-2\varepsilon+w_1-w_2)
\end{aligned} \tag{17}
$$

with $c_1 = -6/5$, $c_2 = -1/2$, $-1/15 < \varepsilon < 3/20$ (to ensure that the contours separate left poles and right poles of Γ functions in the complex plane).

A transformation $w_1 = c_1 + i\,t_1$ and $w_2 = c_2 + i\,t_2$ gives an integral over $\mathbb{R}^2$,

$$
\int_{c_1-i\infty}^{c_1+i\infty} \frac{dw_1}{2\pi i} \int_{c_2-i\infty}^{c_2+i\infty} \frac{dw_2}{2\pi i} \to \frac{1}{4\pi^2} \int_{-\infty}^{\infty} \int_{-\infty}^{\infty} dt_1\, dt_2 \tag{18}
$$

Note that this introduces complex arguments into the Γ-functions of the integrand, e.g., it maps $\Gamma(-1+3\varepsilon-w_1)$ to $\Gamma(-1+3\varepsilon-c_1-i\,t_1)$. Finally a transformation $t_1 = \tan(x_1)$ with $dt_1 =$

Table 8. Results 3-loop UV vertex integral (by DQAGSE×DQAGSE on Mac-Pro in *double* precision); integration err. tolerances $t_r = 10^{-8}$, $t_a = 0$ in each coord. dir.; MAXDIV = 100 (= max. # subdivisions in each coord. direction); $E_r = E_a/|Qf|$ (= integration estim. rel. error); T[s] = Time (elapsed user time) in seconds; $\varepsilon = \varepsilon_\ell = 2^{-\ell}$

3-LOOP VERTEX INTEGRAL $-\Gamma^3(1-\varepsilon)\, I_a^{V3}$ EXTRAPOLATION							
ℓ	E_r	T[s]	RES. C_{-1}	RES. C_0	RES. C_1	RES. C_2	RES. C_3
3	4.6e-10	0.123					
4	7.2e-09	0.065	2.71270658920750307	-124.29196699004			
5	7.1e-09	0.089	-3.26671737757164271	19.214208212662	-765.366267748		
6	6.4e-11	0.103	-2.34303998712649708	-32.511725652267	62.2486740911	-3783.382591	
7	3.2e-10	0.103	-2.40606427439256088	-24.948811180339	-220.100132861	88.82961836	-16521.43
8	7.2e-10	0.105	-2.40408459536474339	-25.439771579237	-180.823300949	-1168.029003	-433.6484
9	1.1e-09	0.101	-2.40411401541469472	-25.424943874060	-183.274814872	-999.9251910	-5454.348
10	1.3e-09	0.100	-2.40411380559384602	-25.425157052041	-183.203187071	-1010.075874	-4804.705
11	1.4e-09	0.100	-2.40411380632043015	-25.425155569821	-183.204191028	-1009.786734	-4842.949
12	1.5e-09	0.100	-2.40411380631917693	-25.425155574937	-183.204184058	-1009.790784	-4841.860
13	1.5e-09	0.100	-2.40411380631918892	-25.425155574852	-183.204184293	-1009.790511	-4842.006
	Exact (14):		-2.40411380631918857	-25.425155574862	-183.204184198	-1009.790681	–

$dx_1/\cos^2(x_1)$, and $t_2 = \tan(x_2)$ with $dt_2 = dx_2/\cos^2(x_2)$ maps $\mathbb{R}^2 \to (-\pi/2, \pi/2) \times (-\pi/2, \pi/2)$. For this example the resulting form is used to generate a sequence of

$$- \Gamma^3(1-\varepsilon)\, I_a^{V3}(\varepsilon) \sim \sum_{k \geq -1} C_k\, \varepsilon^k$$

by a linear extrapolation as $\varepsilon = \varepsilon_\ell \to 0$, to approximate the expansion coefficients as shown in Table 8.

In order to evaluate the (2D) integral numerically, we apply the (one-dimensional) DQAGSE integration program from QUADPACK [2] in both coordinate directions. As part of its regular strategy, DQAGSE performs a global adaptive subdivision and extrapolation with the ϵ-algorithm [7, 8] to treat certain types of end-point singularities. The input parameters of DQAGSE include the requested absolute and relative accuracies, t_a and t_r, respectively, as well as the maximum allowed number of subdivisions, MAXDIV (set as $t_r = 10^{-9}, t_a = 0$, and MAXDIV = 100 for the experiments). Table 8 shows the convergence of the expansion coefficients, as well as the estimated relative error E_r of the outer integration only, and execution time T (in seconds) for each double integration.

5. Conclusions

We presented applications of the PARINT and QUADPACK packages for a direct integration of Feynman loop integrals, with linear extrapolation to compute the expansion coefficients and treat singularities. The methods are fully numerical and viable across the problem types. While the integration strategy is automatic, more work is needed to determine the sequence of ε needed for the extrapolation.

New results using multivariate adaptive integration with PARINT are derived for 3-loop UV-divergent self-energy diagrams with massive internal lines and UV-divergent vertex diagrams with massless propagators. An approach using iterated integration with DQAGSE from QUADPACK combined with linear extrapolation is further shown to provide accurate results based on a 2D Mellin-Barnes representation for a UV-divergent vertex diagram with massless internal lines.

The demonstrated methods provide considerable accuracy within the alotted execution time, compared to other numerical methods that utilize minimal information about the underlying problem. In future work we plan to apply fully numerical methods to various problem classes and widen their domain of applicability for the computation of Feynman loop integrals, for example by the development of suitable transformations to relieve integrand singularities.

Acknowledgments
We acknowledge the support from the National Science Foundation under Award Number 1126438, and the Center for High Performance Computing and Big Data at Western Michigan University. This work is further supported by Grant-in-Aid for Scientific Research (15H03668) of JSPS, and the Large Scale Simulation Program No. 15/16-06 and 16/17-21 of KEK.

Appendix A. C and D functions

Below we list the C and D functions $(C_a^{S3},\ D_a^{S3})$, $(C_b^{S3},\ D_b^{S3})$ of the integrals I_a^{S3}, I_b^{S3}, and $(C_a^{V3},\ D_a^{V3})$, $(C_b^{V3},\ D_b^{V3})$ of the integrals I_a^{V3}, I_b^{V3}, as derived from the corresponding diagrams in Fig 1 and Fig 2. In D_a^{S3} and D_b^{S3} we put $p^2 = 1$ and $m_i^2 = 1$, and in D_a^{V3} and D_b^{V3} the factor $(-p_3^2)$ is not shown.

$$\begin{aligned} C_a^{S3} &= x_{1257}\,x_{3467}\,x_{56} - 2\,x_5\,x_6\,x_7 - x_{3467}\,x_5^2 - x_{1257}\,x_6^2 - x_{56}\,x_7^2 \\ D_a^{S3} &= -x_3^2\,x_5^2 - x_{3467}\,x_{24567}\,x_5^2 + x_{1257}\,x_3^2\,x_{56} + x_1^2\,x_{3467}\,x_{56} + x_{1257}\,x_{3467}\,x_{24567}\,x_{56} \\ &\quad + 2\,x_1\,x_3\,x_7\,x_{56} - x_{24567}\,x_7^2\,x_{56} + 2\,x_1\,x_3\,x_5\,x_6 - 2\,x_{24567}\,x_7\,x_5\,x_6 - x_1^2\,x_6^2 - x_{1257}\,x_{24567}\,x_6^2 \end{aligned} \tag{A.1}$$

$$\begin{aligned} C_b^{S3} &= x_{126}\,x_{4567}\,x_{37} - x_{126}\,x_7^2 - x_6^2\,x_{37} \\ D_b^{S3} &= x_{126}\,x_{37}\,x_4^2 + x_{126}\,x_3^2\,x_{4567} + x_1^2\,x_{37}\,x_{4567} + x_{126}\,x_{2567}\,x_{37}\,x_{4567} + 2\,x_1\,x_{37}\,x_4\,x_6 \\ &\quad - x_3^2\,x_6^2 - x_{2567}\,x_{37}\,x_6^2 + 2\,x_{126}\,x_3\,x_4\,x_7 + 2\,x_1\,x_3\,x_6\,x_7 - x_1^2\,x_7^2 - x_{126}\,x_{2567}\,x_7^2 \end{aligned} \tag{A.2}$$

$$\begin{aligned} C_a^{V3} &= x_{145}\,x_{246}\,x_{356} - 2\,x_4\,x_5\,x_6 - x_{246}\,x_5^2 - x_{356}\,x_4^2 - x_{145}\,x_6^2 \\ D_a^{V3} &= x_1\,x_2\,x_4\,x_{356} + x_1\,x_2\,x_5\,x_6 \end{aligned} \tag{A.3}$$

$$\begin{aligned} C_b^{V3} &= x_{1236}\,x_{1245}\,x_{357} - 2\,x_{12}\,x_3\,x_5 - x_{1245}\,x_3^2 - x_{1236}\,x_5^2 - x_{12}^2\,x_{357} \\ D_b^{V3} &= x_1\,x_2\,x_3\,x_4 + x_1\,x_2\,x_4\,x_5 + x_2\,x_3\,x_4\,x_5 + x_1\,x_2\,x_3\,x_6 + x_1\,x_3\,x_4\,x_6 \\ &\quad + x_1\,x_2\,x_5\,x_6 + x_2\,x_3\,x_5\,x_6 + x_1\,x_4\,x_5\,x_6 + x_3\,x_4\,x_5\,x_6 + x_1\,x_2\,x_3\,x_7 \\ &\quad + x_1\,x_2\,x_4\,x_7 + x_1\,x_2\,x_5\,x_7 + x_2\,x_3\,x_5\,x_7 + x_1\,x_2\,x_6\,x_7 + x_1\,x_4\,x_6\,x_7 \end{aligned} \tag{A.4}$$

References

[1] de Boor C 1971 CADRE: An algorithm for numerical quadrature *Mathematical Software* ed Rice J R (Academic Press, New York) pp 417–449
[2] Piessens R, de Doncker E, Überhuber C W and Kahaner D K 1983 *QUADPACK, A Subroutine Package for Automatic Integration* (*Springer Series in Computational Mathematics* vol 1) (Springer-Verlag)
[3] de Doncker E and Yuasa F 2014 Distributed and multi-core computation of 2-loop integrals doi:10.1088/1742-6596/523/1/012052
[4] Open-MPI http://www.open-mpi.org

IOP Conf. Series: Journal of Physics: Conf. Series **920** (2017) 012005 doi:10.1088/1742-6596/920/1/012005

[5] Berntsen J, Espelid T O and Genz A 1991 *ACM Trans. Math. Softw.* **17** 437–451
[6] Bulirsch R 1964 *Numerische Mathematik* **6** 6–16
[7] Shanks D 1955 *J. Math. and Phys.* **34** 1–42
[8] Wynn P 1956 *Mathematical Tables and Aids to Computing* **10** 91–96
[9] Sidi A 2003 *Practical Extrapolation Methods - Theory and Applications* (Cambridge University Press) iSBN 0-521-66159-5
[10] de Doncker E, Shimizu Y, Fujimoto J, Yuasa F, Cucos L and Van Voorst J 2004 *Nuclear Instruments and Methods in Physics Research A* **539** 269–273 hep-ph/0405098
[11] de Doncker E, Shimizu Y, Fujimoto J and Yuasa F 2004 *Computer Physics Communications* **159** 145–156
[12] Yuasa F, de Doncker E, Hamaguchi N, Ishikawa T, Kato K, Kurihara Y and Shimizu Y 2012 *Computer Physics Communications* **183** 2136–2144
[13] de Doncker E, Fujimoto J, Hamaguchi N, Ishikawa T, Kurihara Y, Shimizu Y and Yuasa F 2011 *Journal of Computational Science (JoCS)* **3** 102–112 doi:10.1016/j.jocs.2011.06.003
[14] Laporta S 2000 *Int. J. Mod. Phys. A* **15** 5087–5159 arXiv:hep-ph/0102033v1
[15] de Doncker E, Yuasa F, Kato K, Ishikawa T, Kapenga J and Olagbemi O 2016 Regularization with numerical extrapolation for finite and UV-divergent multi-loop integrals Submitted
[16] Heinrich G, Huber T and Maître D 2008 *Physics Letters B* **662** 344–352

IOP Conf. Series: Journal of Physics: Conf. Series **920** (2017) 012006 doi:10.1088/1742-6596/920/1/012006

Heterogeneous Hardware Parallelism

Review of the IN2P3 2016 Computing School

Vincent Lafage
Engineer, IPNO, IN2P3, CNRS, 15 rue Georges Clémenceau, 91406 ORSAY Cedex, FR
E-mail: `lafage@ipno.in2p3.fr`

Abstract. Parallel and hybrid Monte Carlo computation.
The Monte Carlo method is the main workhorse for computation of particle physics observables. This paper provides an overview of various HPC technologies that can be used today: multicore (`OpenMP`, `HPX`), manycore (`OpenCL`). The rewrite of a twenty years old Fortran 77 Monte Carlo will illustrate the various programming paradigms in use beyond language implementation. The problem of parallel random number generator will be addressed. We will give a short report of the one week school dedicated to these recent approaches, that took place in École Polytechnique in May 2016.

1. Introduction

It has been over ten years since MOORE's law last provided us with free lunch in computing: the planned, and still observable, growth of the density of transistors per area of chip, was allowing a similar growth of the CPU clock frequencies. This density growth, kept at the constant rate of doubling every two years, has been with us for over forty five years, and while it may disappear even before we reach unmanageable quantum effects in the components at this scale, for financial rather than technical reasons, we have lost the corresponding frequency scaling around 2004. This convenient frequency growth was also correlated with an increasing power dissipation, that lead the industry to the so-called 'heat wall'. The loss of the frequency scaling means that our program have stopped getting faster by simply being ported to a faster CPU. Instead, we have to implicitly rely on the compiler know-how, or worse, explicitly by changing our code to better adapt contemporary architecture. These modern architectures have benefited of the increase in components density in many ways: with wider buses, registers and corresponding larger vector units; with deeper pipelines, better branch prediction and more complex instruction sets; with better use of dead time by sharing multiple logical thread (hyper-threading); with bigger cache memory organized in more levels. Hence the hardware speedup did not stall over these years. But we now have to know better the architecture to get the most out of it. However clever the compilers, we have, at least, to provide it with the options that enable to take advantage of the hardware. And the biggest work remains to provide hints in the code that allows the compiler to make the better use of the computer.

In order to address the recent technologies that appeared and a way to combine them on heterogeneous or hybrid hardware, we decided to prepare the 2016 IN2P3[1] school of computing [1].

[1] french National Institute of Nuclear Physics and Particle Physics

To address various technologies within only one week of school, we needed a short enough program. All the while this program had to be of interest for the researchers of the field, and as such be mostly concentrated on particle physics. What is more, the aim was not to address niceties in parallel programming but more to concentrate on how much raw performance can result from one or the other technology. We decided to elaborate on a simple phenomenology Monte Carlo code, twenty years old. It was designed to compute the Standard Model Quantum Electro Dynamical production of three photons [2] at the LEP collider in CERN as well as the production of the same state resulting from a generic model of new physics. This new physics was modeled with two anomalous coupling of the Z^0 boson to three photons: these dimensional couplings were related to a scale above which their point-like structure would in fact reveal loops of new massive particles such as magnetic monopoles.

Beyond the physics aspects, it was a rather classical code, written in `Fortran` 77 as was traditional in the early nineties. The initially sequential program was thoroughly profiled and checked for potential cache and branch miss.

The first part in the preparatory step was to convert the initial code in various contemporary language. A first translation was of course to `Fortran` 90. `C++`, was beyond doubt a language of choice. `Ada` is a less well known language in high-energy physics, but it allows easy multi-threading embedded in its high-level syntax and was chosen as well. We also had to translate to `C` in view of using `OpenCL`. In fact, this translation work allowed us to see the functional aspect of the initial code, that makes it quite modern.

The next part was to express parallelism at the algorithmic level. The particular choice of the simplest Monte Carlo algorithm lead to what is called an "embarrassingly parallel" program: a problem which can obviously be separated into a number of identical parallel sub-problems, with no, or almost no, dependency or need for communication between those parallel tasks. "Embarrassingly parallel" is a traditional but quite ironically-chosen name, as such a problem proves to be rather "delightfully parallel".

But before going for a parallel version, purely sequential issues must be addressed.

2. Checking for sequential performance

2.1. Profiling

The usual way to access the executable run profile starts with asking the compiler for an instrumented code generation. This extra-code will of course generate some overhead at run time, but this overhead is evenly distributed. This profiling technique is a dynamical one: it requires running the executable, and then use a dedicated application (such as `gprof` [3]) to look at the breakdown of computation time spent.

The resulting report looks like Table 1. One can see some effects of compiler optimization on these report, as exemplified (for the same source code) in Table 2.

2.2. Monitoring cache miss and branch miss

There are further tools to dynamically investigate a code: `valgrind`[2] (see [4]). This program is well-known to help debug memory fault, but can do an even more powerful job of monitoring cache misses and branch misses. We ran our code in this environment, and display the results in Tables 3 and 4. One can see that cache miss is almost non-existent while there is a small, but significant, rate of branch misprediction.

3. Expressing parallelism

This simplest, non adaptive Monte-Carlo algorithm, rely on looping over a basic sampling routine: one first generates an event with a weight (that consists in two steps: a pseudo-random

[2] `http://valgrind.org`

Table 1. Flat profile from execution of an unoptimized sequential version of the code. Each line corresponds to one function of the program.
Column #1 is the percentage of the total running time used by this function.
Column #2 is a running sum of the time (s) accounted for by this function and those above it.
Column #3 is the number of seconds accounted for by this function alone.
Column #4 is the number of times this function was invoked, if it is profiled.
Column #5 is the average time spent (ns) in this function per call, if it is profiled.
Column #6 is the average time spent (ns) in this function and its descendants per call
Column #7 is the name of the function.

% time	cumulative seconds	self seconds	calls	self ns/call	total ns/call	name
12.4	4.44	4.44	495825330	8.95	8.95	gsl_complex_mul [5]
11.2	8.47	4.03	7083219	568.95	2228.41	computeME2 [3]
10.1	12.09	3.62	469162322	7.71	7.71	gsl_complex_mul_real [6]
9.6	15.53	3.44	10000000	344.50	901.05	computeSpinorProducts [4]
5.4	17.48	1.95	256665752	7.60	7.60	gsl_complex_conjugate [10]
5.1	19.29	1.81	10000000	181.00	288.00	generateRambo [7]
4.7	20.96	1.68	200000000	8.38	8.38	gsl_complex_negative [11]
3.7	22.30	1.33				__ieee754_log_avx [12]
2.9	23.34	1.05	42499314	24.71	27.06	gsl_complex_div [15]
2.7	24.30	0.95				__atan_avx [18]
2.3	25.13	0.83	21249657	39.29	118.83	QEDAmplitudePPM [8]
2.3	25.96	0.83	100000000	8.30	8.30	gsl_complex_sub [20]
2.3	26.79	0.83				scalbn [21]
2.1	27.52	0.73	21249657	34.59	114.13	QEDAmplitudePMM [9]
1.8	28.16	0.64	10000000	64.00	82.52	computeScalarProducts [22]
1.6	28.75	0.58	120000000	4.83	4.83	ocl_rng_rand48_nextState [23]
...						
0.9	34.55	0.34	10000000	34.00	34.00	sortPhotons [32]
0.8	34.85	0.30	7083219	42.35	126.15	anomalousAmplitudePPP [19]
0.5	35.04	0.19	28332876	6.71	6.71	gsl_complex_add [33]

Table 2. Profile from executable optimized with `-O3 -march=native -funroll-loops`

% time	cumulative seconds	self seconds	calls	self ns/call	total ns/call	name
24.3	4.69	4.69	469162322	10.00	10.00	gsl_complex_mul_real [3]
7.3	6.10	1.41	100000000	14.10	14.10	gsl_complex_sub [5]
7.3	7.51	1.41				__ieee754_log_avx [7]
7.3	8.92	1.41				sincos [6]
7.1	10.28	1.36	200000000	6.80	6.80	gsl_complex_negative [8]
5.6	11.36	1.08				scalbn [9]
5.0	12.33	0.97				computeSpinorProducts [1]
4.4	13.18	0.85				generateRambo [4]
4.3	14.00	0.82				computeME2 [2]
3.2	14.61	0.61	120000000	5.08	5.08	ocl_rng_rand48_getDouble [11]
2.6	15.12	0.51	495825330	1.03	1.03	gsl_complex_mul [13]
2.5	15.61	0.49				__scalbn [14]
2.1	16.01	0.40				bsloww [15]
1.8	16.36	0.35	28332876	12.35	12.35	gsl_complex_add [17]
1.8	16.70	0.34				selectEvent [19]
1.7	17.03	0.33	42499314	7.76	8.24	gsl_complex_div [16]
...						
1.3	18.14	0.26	269997256	0.96	0.96	gsl_complex_abs2 [26]
1.0	18.33	0.19	256665752	0.74	0.74	gsl_complex_conjugate [28]
0.8	18.48	0.15				computeScalarProducts [27]
0.7	18.61	0.13	21249657	6.12	29.49	QEDAmplitudePMM [10]

IOP Conf. Series: Journal of Physics: Conf. Series **920** (2017) 012006 doi:10.1088/1742-6596/920/1/012006

Table 3. Cache & branch misses
These listings describe the number of fetch (refs) of instruction (I) and data (D) from cache Level 1 (I1, D1) or Last Level (LLi, LLd, total in LL), as well as the number of cache-miss occurring and the corresponding rate. In the case of data, the number of read access and write access are shown separately. The number of branches occurring in the execution, and the number of branch mispredictions are also displayed.

```
time valgrind --tool=cachegrind --branch-sim=yes mc
...
==32423==
==32423== I   refs:      21,866,876,065
==32423== I1  misses:            90,912
==32423== LLi misses:             2,966
==32423== I1  miss rate:           0.00%                                                      => 4,2.10^-6
==32423== LLi miss rate:           0.00%
==32423==
==32423== D   refs:      10,907,321,258  (7,120,262,982 rd   + 3,787,058,276 wr)
==32423== D1  misses:            17,778  (       15,391 rd   +         2,387 wr)
==32423== LLd misses:             9,880  (        8,407 rd   +         1,473 wr)
==32423== D1  miss rate:            0.0% (          0.0%     +           0.0%  )                => 1,6.10^-6
==32423== LLd miss rate:            0.0% (          0.0%     +           0.0%  )
==32423==
==32423== LL refs:              108,690  (      106,303 rd   +         2,387 wr)
==32423== LL misses:             12,846  (       11,373 rd   +         1,473 wr)
==32423== LL miss rate:             0.0% (          0.0%     +           0.0%  )
==32423==
==32423== Branches:         920,763,441  (  853,502,037 cond +    67,261,404 ind)
==32423== Mispredicts:       48,791,446  (   42,788,533 cond +     6,002,913 ind)
==32423== Mispred rate:             5.2% (          5.0%     +           8.9%   )
```

Table 4. Cache & branch misses, optimized with `-O3 -march=native -funroll-loops`

```
 Elapsed time per event           : 1.87726e-06         => -1/3
...
time valgrind --tool=cachegrind --branch-sim=yes mc
...
==20432== I   refs:      8,807,755,982
==20432== I1  misses:           16,437
==20432== LLi misses:            2,828
==20432== I1  miss rate:          0.00%
==20432== LLi miss rate:          0.00%
==20432==
==20432== D   refs:      3,307,797,849  (1,911,804,779 rd   + 1,395,993,070 wr)
==20432== D1  misses:           17,650  (       15,274 rd   +         2,376 wr)
==20432== LLd misses:            9,847  (        8,376 rd   +         1,471 wr)
==20432== D1  miss rate:           0.0% (          0.0%     +           0.0%  )
==20432== LLd miss rate:           0.0% (          0.0%     +           0.0%  )
==20432==
==20432== LL refs:              34,087  (       31,711 rd   +         2,376 wr)
==20432== LL misses:            12,675  (       11,204 rd   +         1,471 wr)
==20432== LL miss rate:            0.0% (          0.0%     +           0.0%  )
==20432==
==20432== Branches:        701,657,379  (  642,396,799 cond +    59,260,580 ind)
==20432== Mispredicts:      30,876,538  (   30,873,904 cond +         2,634 ind)
==20432== Mispred rate:            4.4% (          4.8%     +           0.0%   )
```

number generator, such as the one in [5], and a routine to translate these to a random event, such as [6]). The program then checks if it passes cuts, computes matrix element, weights it, sums it up, and starts again. As each event is supposed to be independent, and there are no explicit dependencies, the algorithm should lend itself to parallelization. The parallel threads only have to meet at the end of the loop to sum up their partial results, an operation known as reduction.

3.1. Random number generation

Even good generators, such as the one from KNUTH [7, 8], that avoid the classical problem of linear congruential generator (see [9]) are essentially sequential: you have to go through all the sequence to reach a given index, and this is the major obstacle to parallelizing the algorithm. The famous "Mersenne twister" generator [10], allows, among others, to "fast forward" efficiently through the sequence, and having a very long period, can start by generating the seeds of well separated sub-sequences, one for each thread. So doing keeps the sequential nature of the generator, but allows to have non-overlapping parallel sub-sequences. This was our starting point for this school. But since then we have been introduced to `Random123`, a random generator that is a direct function of the index of the sequence (see [11] in [12], as well as [13]). What is more, it is blessed with very good randomness properties, passing the `BigCrush` [14] test suite, and we are porting it to all future versions of our code.

4. Technologies illustrated

4.1. OpenMP

This technology has already been with us for almost twenty years. It should, in our opinion, be considered as the natural entrance for beginner and non-expert wishing to turn their old software into a multithreading package. It relies on compiler directives that are commented by default but can be turned on when parallelizing. It allows to keep the purely sequential version alive within the same code base. Parallelization is mostly an industrialization process: it is about organizing data space so that different actors can process independently, sharing only the minimal requested information. `OpenMP` directive allow to specify this data and process organization.

We should now discuss the scaling of the performance and what it tells us about the underlying architecture and, paradoxically, the real sequential performance. The program has been run on a 16 physical cores machine with hyper-threading, with setting the number of simultaneous thread from 1 to 32 (see Figures 1 and 2).

It turns out that on 32 logical cores, the CPU usage was about 27: a result accepted with mixed feelings, as it was impressive, but yet not perfect, and with the program so cleanly parallel in principle, the cause of the default in scaling was not clear. What is more, the speedup, ideal up to 11 threads, has a plateau until 15 threads, before increasing again more and more slowly.

In order not to be fooled by the speedup, we need to know how close to the full potential of the hardware we could reach. This is now possible with the recent tool `maqao`[3] (see [15]) a *Modular Assembly Quality Analyzer and Optimizer* which can be used for static *binary executable* analysis. This reports as exemplified in Table 5 and shows that at most the sequential version was using but 13% of the computing capability of the CPU. In fact, the program did not express vectorization (see Table 5), and the vector arithmetic unit is left unsupplied. This point shed some light on our understanding of the effect of hyper-threading on a code that in principle should absolutely not benefit from it: there is no input/output in the loop and hardly any cache-miss, be it for the few data required or for the instructions. The alternate thread sharing a given physical core should then have almost no opportunity to use dead time, and so be almost useless.

As the speedup is quietly saturating (at about 17.6), but above the number of physical cores (16), we can see that some ten percents of hyper-threading occurs, and points to some dead time

[3] `http://www.maqao.org/`

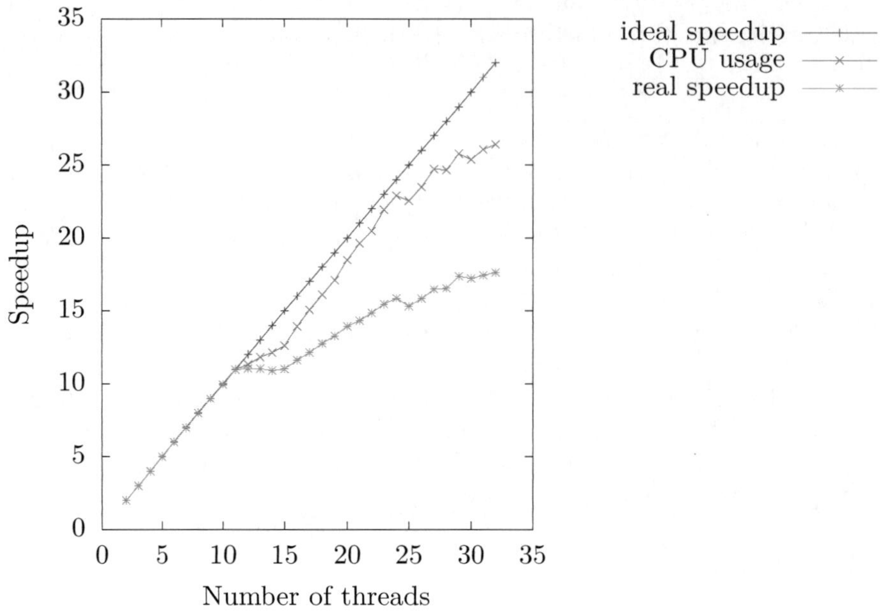

Figure 1. Multi-thread speedup

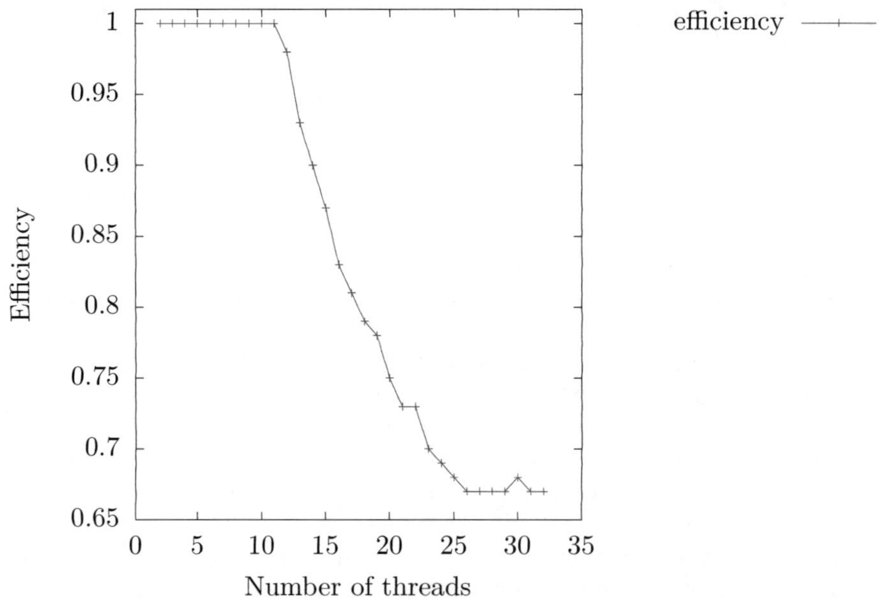

Figure 2. Efficiency

Table 5. Excerpt of maqao output example
This program diagnoses various types of inefficiencies directly in the executable code of one of the function in the program: here, the function bppp computing one of the matrix element. Diagnosis is reported in paragraphs, suggesting improvements for each problem. In this case, the report exhibits issues with a lack of vectorization of the code.

```
$ maqao.intel64 cqa mc fct-body=bppp
Section 1: Function: bppp
========================

...
13\% of peak computational performance is used (2.13 out of 16.00 FLOP per cycle (5.97 GFLOPS @ 2.80GHz))

Code clean check
----------------
Detected a slowdown caused by scalar integer instructions (typically used for address computation).
By removing them, you can lower the cost of an iteration from 23.00 to 17.00 cycles (1.35x speedup).
...

Vectorization status
--------------------
Your function is probably not vectorized (store and arithmetical SSE/AVX instructions are used in scalar mode and,
for others, at least one is in vector mode).
Only 26\% of vector length is used.

Vectorization
-------------
Your function is processing FP elements but is NOT OR PARTIALLY VECTORIZED and could benefit from full vectorization.
By fully vectorizing your function, you can lower the cost of an iteration from 23.00 to 5.75 cycles (4.00x speedup).
Since your execution units are vector units, only a fully vectorized function can use their full power.

Two propositions:
 - Try another compiler or update/tune your current one:
 - Make array accesses unit-stride.
  * If your function streams arrays of structures (AoS), try to use structures of arrays instead (SoA):
do i a(i)%x = b(i)%x (slow, non stride 1) => do i a%x(i) = b%x(i) (fast, stride 1)
...

FMA
---
Presence of both ADD/SUB and MUL operations.

 - Try to change order in which elements are evaluated (using parentheses) in arithmetic expressions containing
both ADD/SUB and MUL operations to enable your compiler to generate FMA instructions wherever possible.
```

in the sequential version. It would be interesting to track the excess of CPU usage appearing above 12 threads.

As far as precision is concerned, runs with 1 to 32 threads lead to the same result within a relative standard deviation of 3.8×10^{-9}, which reflects different rounding errors depending on summation order. This is negligible in a multi-core approach, but with a much larger number of threads, as on GPU, it might be interesting to implement a parallel KAHAN summation algorithm [16].

4.2. Ada tasking

Amongst the first attempts at writing a parallel version of the Monte Carlo code, Ada was chosen for its ability for concurrent programming. The sequential version did not suffer more than 20 percent speed penalty compared to the Fortran version. But without a good parallel pseudo-random number generator, one task was dedicated to producing random number that would be used by other tasks, limiting the gullibility. The next step will use the Random123 generator.

4.3. The `Go` programming language

A sequential version of the Monte Carlo code was written in Google's `Go` language `clang.org` achieving the same level of performance as `Fortran`. After implementing the `Random123` generator, the concurrent aspect will be investigated.

4.4. MPI

Message Passing Interface (`MPI`) is among the oldest (1994) common parallel, and more specifically distributed memory, technologies. It allows for a large and uniform scalability (as one can always extend the number of node, as long as the problem can cope with this extension). It relies on the conventional communicating sequential processes (CSP) [17] model. It has been used both independently and as one component of "hybridization" in the GPU approach of the school: `MPI+OpenCL`

4.5. GPU with `OpenCL`[4]

General-Purpose computing on Graphics Processing Units (GPGPU) are a set of frameworks to accelerate computation by exploiting the massive vectorization of Graphics Processing Units. For instance `CUDA`, `OpenACC` and `OpenCL`. These last two are programming standards.

`OpenCL` [18] is a programming language based on `C99`, extended to facilitate use of parallelism with vector types and operations, synchronization, and functions that allows to write both CPU programs and GPU executed kernels. `OpenCL` can also run kernels on CPU. It has been used both independently and as one component of "hybridization" in the GPU approach of the school: `MPI+OpenCL`.

The performance was tested on the GridCL platform [19] built of four node servers (each with a dual processor Intel E52650 at 2 GHz, 16 cores, 64 GB of RAM, with AVX vectorization on four doubles, and hyperthreading 32 threads) connected with Infiniband. Each node is equipped with two GPU boards: two nodes with NVidia K20[5] and the two others with Intel Xeon Phi[6]. Another standalone server is equipped with six NVidia Titan GPU boards. Table 6 details the speedup observed when comparing various configuration and shows an impressive factor 147 when using simultaneously the six NVidia Titan GPU boards.

Table 6. Measurement of run times speed-up compared to the sequential `C++`.
The line "One Device" displays the speedup when only one GPU board (or CPU set of cores) is used, the line "All Devices" when all GPU boards of the same kind are used (across Infiniband where applicable).

	C++ 1-core CPU	**OpenCL** 16-cores CPU	**OpenCL** K20 NVidia GPU	**OpenCL** Titan NVidia GPU	**OpenCL** Xeon Phi Intel GPU
One Device	1	24.7	22.1	26.7	16.4
All Devices	1		63.0	146.7	47.5

4.6. `HPX` libraries[7] [20]

`HPX` (High Performance ParalleX) is a `C++` library implementing the `ParalleX` execution model proposed to go beyond CSP model [17].

[4] `https://www.khronos.org/opencl/`
[5] `http://international.download.nvidia.com/pdf/kepler/NVIDIA-Kepler-GK110-GK210-Architecture-Whitepaper.pdf`
[6] `https://software.intel.com/en-us/articles/intel-xeon-phi-coprocessor-codename-knights-corner`
[7] `http://stellar-group.org/libraries/hpx/`

The goal of the ParalleX model of computation is to address the key challenges of efficiency, scalability, sustained performance, and power consumption with respect to the limitations of conventional programming practices (e.g., MPI), by enabling a new computing dynamic through the application of message-driven computation in a global address space context with lightweight synchronization.

5. Other presented technologies

5.1. DSL: Domain Specialized Languages

Although we did not implement such a solution for our simple Monte Carlo with DSL, the problem we met in parallelizing helps us understanding the approach. Often a program gives a sequential solution to a problem relying on nests of loops, that in turn get across the specific architecture and / or memory model of the parallel computing environment. A complete rewriting of the code then seems unavoidable, and we then understand that implementing a solution has been losing the generality of the problem: we then call for a more abstract expression of our problem. To fill the gap of abstraction between our mathematical vision of the problem and the actual code, we need a higher-level language (closer tho the mathematical language) and a corresponding translator to our fast computing language (`Fortran`, `C++`) that will transfer formula to vectorized expression or correctly nested parallel loops in a much faster and more versatile way than we can code "by hand". We can see this as a dedicated code generator.

The `Loopy` system (coded in `Python`) is such a generator and was presented by its author [21].

5.2. OpenCL for FPGA

The general idea of using a dedicated hardware acceleration involves a high cost. Field Programmable Gate Array (FPGA) aims at combining the speed of dedicated hardware with the flexibility of programming. But in practice, "compiling a hardware solution" is lengthy and delicate and requires a range of expertise. Building a dedicated hardware allows to exploit both larger and/or more adequate vector computing units with way deeper computation pipeline.

The idea here is to generate the specific architecture directly from an `OpenCL` code. However faster it is, the development cycle is still affected by large compilation/synthesizing time (order of many hour to one day).

6. Conclusion

Sixty persons applied to attend the school for thirty places. The goal of this school was to exemplify ways to turn a well-tested sequential problem into one exploiting most of the performance of contemporary computers. And it succeeded in achieving impressive performance such as speedup of 147 on a multi-GPU nodes. Hybrid programming such as `MPI+OpenCL` was also up to the task for large scalability.

A next edition would certainly emphasize other approaches such as `CUDA Fortran` [22] or the promising `OpenACC` [23] that take the same approach "by compiler directives" as `OpenMP` to turn a sequential application in one using GPU. Another environment of interest, `Kokkos`, implements a programming model in standard C++ for writing performance portable applications [24, 25].

As the goal is to help users move from the sequential to the parallel paradigm, tools like `kremlin` (see [26, 27]) should be investigated.

References

[1] Chamont D L, Grasseau G L and Lafage V I 2016 École Informatique IN2P3 « Parallélisme sur Matériel Hétérogène » URL `https://indico.in2p3.fr/event/13126/`
[2] Baillargeon M, Boudjema F, Chopin E and Lafage V 1996 *Z. Phys.* **C71** 431–442 (*Preprint* `hep-ph/9506396`) `doi:10.1007/BF02907001`

[3] Graham S L, Kessler P B and McKusick M K 2004 *SIGPLAN Not.* **39** 49–57 ISSN 0362-1340 URL http://doi.acm.org/10.1145/989393.989401 doi:10.1145/989393.989401
[4] Nethercote N and Seward J 2007 *SIGPLAN Not.* **42** 89–100 ISSN 0362-1340 URL http://doi.acm.org/10.1145/1273442.1250746 doi:10.1145/1273442.1250746
[5] Knuth D E 1997 *The Art of Computer Programming, Volume 2 (3rd Ed.): Seminumerical Algorithms* (3rd ed.) ed vol 2 (Boston, MA, USA: Addison-Wesley Longman Publishing Co., Inc.) ISBN 0-201-89684-2
[6] Kleiss R, Stirling W J and Ellis S D 1986 *Comput. Phys. Commun.* **40** 359 rA(ndom) M(omenta) B(eautifully) O(rganized), a democratic multi-particle phase space generator. doi:10.1016/0010-4655(86)90119-0
[7] Knuth D E 1969 *The Art of Computer Programming: Seminumerical Algorithms, Chapter 3. Random Numbers* chap 3. Random Numbers vol 2 of Knuth:1969:ACP:270146 [8] (1st ed.) ed
[8] Knuth D E 1969 *The Art of Computer Programming, Volume 2 (1st Ed.): Seminumerical Algorithms* (1st ed.) ed vol 2 (Boston, MA, USA: Addison-Wesley Longman Publishing Co., Inc.) ISBN 0-201-03802-1
[9] Marsaglia G 1968 *Proceedings of the National Academy of Sciences of the United States of America* **61** 25–28 ISSN 00278424 URL http://www.jstor.org/stable/58853 doi:10.1073/pnas.61.1.25
[10] Matsumoto M and Nishimura T 1998 *ACM Trans. Model. Comput. Simul.* **8** 3–30 ISSN 1049-3301 URL http://doi.acm.org/10.1145/272991.272995 doi:10.1145/272991.272995
[11] Salmon J K, Moraes M A, Dror R O and Shaw D E 2011 Parallel random numbers: As easy as 1, 2, 3 in Lathrop:2011:2063384 [12] pp 16:1–16:12 URL http://doi.acm.org/10.1145/2063384.2063405
[12] 2011 *SC '11: Proceedings of 2011 International Conference for High Performance Computing, Networking, Storage and Analysis* (New York, NY, USA: ACM) ISBN 978-1-4503-0771-0
[13] 2011 URL http://www.deshawresearch.com/resources_random123.html
[14] L'Ecuyer P and Simard R 2007 *ACM Trans. Math. Softw.* **33** 22:1–22:40 ISSN 0098-3500 doi:10.1145/1268776.1268777
[15] Djoudi L, Noudohouenou J and Jalby W 2008 The Design and Architecture of MAQAOAdvisor: A Live Tuning Guide *Proceedings of the 15th International Conference on High Performance Computing* HiPC'08 (Berlin, Heidelberg: Springer-Verlag) pp 42–56 ISBN 978-3-540-89893-1 URL http://dl.acm.org/citation.cfm?id=1791889.1791899
[16] Kahan W 1965 *Commun. ACM* **8** 40– ISSN 0001-0782 URL http://doi.acm.org/10.1145/363707.363723 doi:10.1145/363707.363723
[17] Hoare C A R 1978 *Commun. ACM* **21** 666–677 ISSN 0001-0782 URL http://doi.acm.org/10.1145/359576.359585 doi:10.1145/359576.359585
[18] Munshi A 2009 The opencl specification *2009 IEEE Hot Chips 21 Symposium (HCS)* pp 1–314
[19] 2013 URL http://polywww.in2p3.fr/projet-gridcl
[20] Tabbal A, Anderson M, Brodowicz M, Kaiser H and Sterling T 2011 *SIGMETRICS Perform. Eval. Rev.* **38** 81–87 ISSN 0163-5999 URL http://doi.acm.org/10.1145/1964218.1964232 doi:10.1145/1964218.1964232
[21] Klöckner A 2014 Loo.py: transformation-based code generation for GPUs and CPUs *Proceedings of ARRAY '14: ACM SIGPLAN Workshop on Libraries, Languages, and Compilers for Array Programming* (Edinburgh, Scotland.: Association for Computing Machinery)
[22] NVIDIA and the Portland Group 2016 Cuda fortran programming guide and reference. URL https://www.pgroup.com/doc/pgi16cudaforug.pdf
[23] CAPS Enterprise, Cray Inc, NVIDIA and the Portland Group 2013 The openacc application programming interface, v2.0 URL https://www.openacc.org/sites/default/files/inline-files/OpenACC_2_0_specification.pdf
[24] Edwards H C and Sunderland D 2012 Kokkos array performance-portable manycore programming model *Proceedings of the 2012 International Workshop on Programming Models and Applications for Multicores and Manycores* PMAM '12 (New York, NY, USA: ACM) pp 1–10 ISBN 978-1-4503-1211-0 URL http://doi.acm.org/10.1145/2141702.2141703
[25] Edwards H C and Trott C 2015 Kokkos, Manycore Device Performance Portability for C++ HPC Applications URL http://on-demand.gputechconf.com/gtc/2015/presentation/S5166-H-Carter-Edwards.pdf
[26] 2011 URL http://kremlin.ucsd.edu
[27] Garcia S, Jeon D, Louie C M and Taylor M B 2011 Kremlin: Rethinking and rebooting gprof for the multicore age *Proceedings of the 32Nd ACM SIGPLAN Conference on Programming Language Design and Implementation* PLDI '11 (New York, NY, USA: ACM) pp 458–469 ISBN 978-1-4503-0663-8 URL http://doi.acm.org/10.1145/1993498.1993553

IOP Conf. Series: Journal of Physics: Conf. Series **920** (2017) 012007 doi:10.1088/1742-6596/920/1/012007

CompHEP: developments and applications

E.E. Boos, V.E. Bunichev, M.N. Dubinin, V.A. Ilyin and V.I. Savrin
(on behalf of CompHEP Collaboration)

Skobeltsyn Institute of Nuclear Physics, Lomonosov Moscow State University, Leninskiye Gory, 119991 Moscow, Russian Federation

E-mail: `comphep@theory.sinp.msu.ru`

Abstract. New developments of the CompHEP package and its applications to the top quark and the Higgs boson physics at the LHC collider are reviewed. These developments were motivated mainly by the needs of experimental searches of DO (Tevatron) and CMS (LHC) collaborations where identification of the top quark and the Higgs boson in the framework of the Standard Model (SM) or possible extensions of the SM played an important role. New useful features of the CompHEP Graphics User Interface (GUI) are described.

1. Introduction

The general plan of development of the CompHEP project was proposed more than twenty five years ago [1]. Despite the fact that the initial objectives of the project were related to the calculation of the most relevant for the 1990-ies signal and background processes at the e^+e^- collider LEP2 and the $p\bar{p}$ collider Tevatron, very general approach to the simulation of particle production channels was developed. It was supposed to generate unweighed events for the detector simulation of any signal in the SM or beyond using a database for basic Lagrangians which would allow to generate complete gauge-invariant sets of diagrams for the signal and the irreducible background in the fully automatic regime with the following evaluation of squared amplitudes. The original scheme can be found in [2]. The primary version of CompHEP used the Pascal language for PC AT and generated automatically REDUCE [3] format code for the squared amplitude corresponding to a gauge-invariant set of diagrams [2, 4]. It was expected that time that analytic results for the differential and the total cross sections could be straightforwardly obtained and used to interpret an experimental data. Obviously, such a development strategy did not allow to reduce to an acceptable form bulky analytical expressions for the matrix element squared which correspond to several tens or hundreds of diagrams, forming a complete set. So developers had to give up modelling based on compact analytical results (such as signal processes library in the approximation of infinitely small width) and go to generation of FORTRAN codes on the basis of exact symbolic matrix element squared, which were used in the following numerical integration. Automatic generation of FORTRAN codes suitable for the following Monte Carlo (MC) integration by means of BASES generator [5, 6], which was provided by the GRACE [7] group, was an important step forward. Jointly, although using different methods, MC calculations were performed for a large set of LEP2 processes [8], which allowed to test reliably all the algorithms involved in calculation. A number of phenomenological applications were related not only to e^+e^- mode, but also to possible γe and $\gamma\gamma$ modes of JLC and TESLA linear colliders [9]. In the following the channels of four-fermion production at LEP2 were evaluated [10]–[14], which were the first calculations of high precision at the complete tree-level. Numerous cross-checks of CompHEP and GRACE algorithms and comparisons of results for a large set of physical channels were very important for the following applications to collider physics.

In the last twenty years extensive symbolic and numerical calculations and generation of unweighed events for LEP2, HERA, TESLA, Tevatron, ILC and LHC analyses [15]–[30],[31] have been performed by means of CompHEP versions 3 and 4 [32, 33, 34, 35]. Single top quark production at the Tevatron $p\bar{p}$ collider [36] was followed by the single top observation at the LHC [37, 38]. CompHEP-based calculations of the complete tree-level sets of diagrams and the following event generation at the next-to-leading order (NLO) level were used for analyses of the single top quark signal [39]. Simulations of the single top quark production are important both for precise measurements and for searches of new physics [40, 41]. Convincing statistical significance determined a discovery of the Higgs-like boson signal at the LHC in 2012 [42, 43], which was investigated intensively in many production channels for identification purposes, either it was a signal of the Standard Model (SM) Higgs boson or a signal beyond the SM (BSM signal). Simulation of the Higgs boson signal at the LHC can be sensibly performed only for complete sets of tree-level diagrams, when large backgrounds, typical for hadronic colliders, can be put under control. A number of CompHEP technical developments which have been used in the LHC analyses are described in some details below.

2. Model-independent SM extensions for the top quark and the Higgs boson production

In order to measure precisely the couplings of the Higgs boson to vector bosons and fermions of the SM or a model BSM, a combined analysis of all the observed production channels has been performed. Combined analysis gives a possibility to extract 'pseudo observables' [44, 45] which depend on ratios of the experimentally observed production cross sections and decay widths to those calculated in the SM. In such analyses it is assumed that properties of the SM Higgs boson are modified by a new physics at the TeV scale Λ, so the SM is an effective theory at the $W^\pm, Z$ scale. In the framework of the technical modification under consideration either a generic set of the higher-dimension effective operators is introduced in addition to the SM Lagrangian (note that the same set of effective operators can be a remnant of different field theory models at a multiTeV energy scale), or consequences of a definite renormalizable model are evaluated at the electroweak scale. In both types of extensions the question to what extent the observed boson is consistent with the SM Higgs boson seems most interesting to be addressed.

Extensions by the dimension-six effective operators have been considered recently in different bases (e.g.[46]-[49]). In the following we use [50, 51, 52] where a rather general framework of the SM extension by the dimension-six effective operators was analysed. At the electroweak energy scale the new physics degrees of freedom can be integrated out and the low-energy effective Lagrangian can be written as

$$L_{eff} \quad = \quad L_{SM} + \frac{1}{\Lambda^2} \sum_{k=V,F} C_{k\Phi} O_{k\Phi}$$

where Λ is a scale of new physics and the anomalous couplings $C_{k\Phi}$ modify the SM Higgs boson couplings to the vector bosons and to the fermions. A general set of the dimension-six operators from [50] in the Buchmueller-Wyler basis [53] modified by the subtraction of v.e.v.: $\Phi^\dagger\Phi \to \Phi^\dagger\Phi - v^2/2$ (see [54]) to avoid undesirable mixing in the gauge field kinetic terms, can be reduced to a restricted set of only five fermion-Higgs and vector boson-Higgs operators $O_{t\Phi}$, $O_{b\Phi}$, $O_{\tau\Phi}$ and $O_\Phi^{(1)}$, $O_{\Phi G}$, disposing the tensor structure of interaction vertices identical to the

SM and dependent only on the two anomalous couplings. The coupling c_V rescales the vector boson-Higgs vertices, and the coupling c_F rescales the fermion-anti-fermion-Higgs vertices. The anomalous couplings C_n, C_{mn} in front of the dimension-six operators O_n, O_{mn} are conformably redefined [50], for example

$$c_F = 1 + C_{t\Phi} \cdot \frac{v^2}{\Lambda^2}, \quad c_V = 1 + \frac{v^2}{2\Lambda^2} \cdot C_{\Phi}^{(1)}, \quad c_G = c_F + \frac{6\pi}{\alpha_s} \cdot C_{\Phi G} \cdot \frac{v^2}{\Lambda^2}, \quad \ldots$$

(here v =246 GeV). The last equation above means that in the case under consideration the one-loop $H \to \gamma\gamma$ and $H \to gg$ vertices are 'resolved' in the sense that the anomalous parameters c_V and c_F are included in the one-loop effective c_G and c_γ couplings which parametrise $H \to gg$ and $H \to \gamma\gamma$ vertices. This circumstance is different from 'kappa-framework'.

Anomalous interactions of the third generation fermions (t,b) appear from the general local $SU(3) \times SU(2) \times U(1)$ invariant effective Lagrangian terms [56]. Seven dimension-six effective operators provide anomalous contributions to the Wtb vertex. Various contributions of these operators which include the gauge bosons and the Higgs sector isodoublet are usually denoted by O_{qW}, $O^3_{\Phi q}$, O_{Db}, $O_{bW\Phi}$, O_{Dt}, $O_{tW\Phi}$ and O_{t3}. Experimental restrictions on the rare processes exclude some contributions. Application of the equations of motion move some operators to other category leaving only two meaningful operators of the seven, $O_{bW\Phi}$ and $O_{tW\Phi}$:

$$\begin{aligned} O_{tW\Phi} &= C_{tW\Phi}[(\bar{q}_L\sigma^{\mu\nu}\tau^i t_R)\Phi + \Phi^+(\bar{t}_R\sigma^{\mu\nu}\tau^i q_L)]W^i_{\mu\nu} \\ O_{bW\Phi} &= C_{bW\Phi}[(\bar{q}_L\sigma^{\mu\nu}\tau^i b_R)\Phi + \Phi^+(\bar{b}_R\sigma^{\mu\nu}\tau^i q_L)]W^i_{\mu\nu} \end{aligned}$$

where q_L is the left-handed third-family doublet, Φ is the Higgs boson doublet, $\tau^i = \sigma^i/2$, $W_\mu = \tau^i W^i_\mu$. Redefining anomalous couplings C after rotation to the physical fields $f_{2L} = \frac{C_{tW\Phi}}{\Lambda^2}\frac{v\sqrt{2}m_W}{g}$ and $f_{2R} = \frac{C_{bW\Phi}}{\Lambda^2}\frac{v\sqrt{2}m_W}{g}$ ($v\sqrt{g^2+g'^2} = 2m_Z$, $s^2_W = 1 - m^2_W/m^2_Z$) we get the standard representation of the anomalous top quark interaction Lagrangian

$$L_{Wtb} = \frac{g}{\sqrt{2}}\bar{b}\gamma^\mu(f_{1L}P_L + f_{1R}P_R)t\,W^-_\mu + \frac{g}{\sqrt{2}}\bar{b}\frac{\sigma^{\mu\nu}}{m_W}(f_{2L}P_L + f_{2R}P_R)t\,W^-_{\mu\nu} + h.c. \quad (1)$$

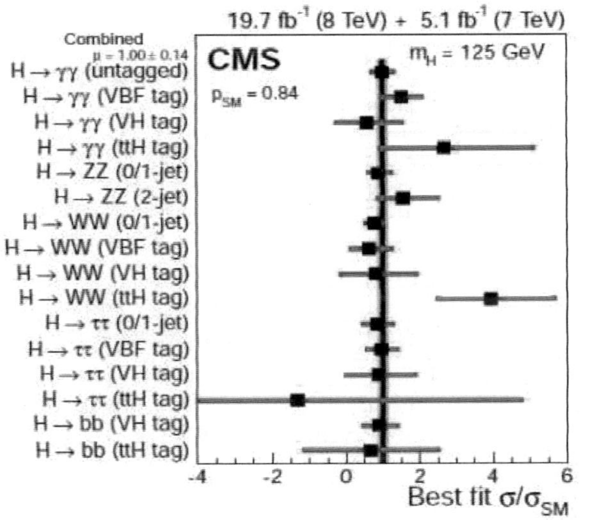

Figure 1. Signal strength and signal strength error for various groups of production channels. The best-fit σ/σ_{SM} for the overall combined analysis where combinations of channels grouped by the production mode and specific kinematics in the final state taken from [55] is indicated by vertical line. Horizontal bars indicate $\pm 1\sigma$ uncertainties in the best-fit σ/σ_{SM} values for individual channels, both statistical and systematic uncertainties are included. Combinations of channels are grouped by dominant decay modes when more than one decay mode contributes to the same final state. Tags in brackets indicate a specific production mechanism.

3. Global fits for the Higgs boson production channels

The method of exclusion contours reconstruction [57, 58, 59, 60] in the anomalous parameter space has been introduced to CompHEP version 4. Theoretical signal strengths in the infinitely small width approximation (ISW or production × decay approximation) and for the complete gauge-invariant sets of diagrams are defined as

$$\mu_i^{ISW} = \frac{[\sum_j \sigma_{j\to h} Br(h\to i)]_{SME}}{[\sum_j \sigma_{j\to h} Br(h\to i)]_{SM}}; \qquad \mu_i = \frac{\left[\sum_{j=1}^{N_{ch}} \sigma_{j\to H(off-shell)\to i}\right]_{SME}}{\left[\sum_{j=1}^{N_{ch}} \sigma_{j\to H(off-shell)\to i}\right]_{SM}}$$

where for ISW i is the number of Higgs boson decay channel and j is the number of Higgs production process for a given final state in the SM and in the SM extension (SME). If a signal strength $\hat{\mu}_i$ for an individual channel (see Fig.1) can be expressed using the observed cross section σ_{obs}, the background cross section σ_{backgr} and the SM signal cross section σ^{SM}_{signal}, then the global χ^2 for the signal strengths $\hat{\mu}_i$ is defined as

$$\hat{\mu}_i = \frac{\sigma_{obs,i} - \sigma_{backgr,i}}{\sigma^{SM}_{signal,i}}; \qquad \chi^2(\mu_i) = \sum_i^{N_{ch}} \frac{(\mu_i - \hat{\mu}_i)^2}{\sigma_i^2}$$

for the number of production channels N_{ch}. Minimization of $\chi^2 \to \chi^2_{min}$ gives us 1σ, 2σ and 3σ regions $\chi^2 = \chi^2_{min} + \Delta\chi^2$ where $\Delta\chi^2$ is defined by cumulative distribution function. Assuming that the signal strengths of various channels have Gaussian distributions with the probability density functions (PDF's) having the expected values $\hat{\mu}_i$ and the dispersions σ_i normalized to one, combined PDF for a number of production channels can be found by multiplication of PDF's for the individual channels. The combined probability density function is also Gaussian characterized by μ_c and σ_c, $1/\sigma_c^2 = \sum_i^{N_{ch}} \cdot 1/\sigma_i^2$. The combined PDF allows one to determine, for example, 95% CL exclusion upper μ_U and lower μ_L limits on the signal strength parameter integrating the combined pdf from $\hat{\mu}$ to μ_U and from μ_L to $\hat{\mu}$, respectively, then equating the result to 0.95/2. If the SM is fully adequate, the values of μ_i are as close to one as allowed by experimental errors. Calculation of the $\Delta\chi^2$, see Fig.2, for the best fit defines a *given number* × *CL* contours corresponding to the departure of the SM point $(1, 1)$ from the best fit point in the (c_V, c_F) parameter plane. Contours in Fig.3 correspond to 65%, 90% and 99% best fit CL regions with $\Delta\chi^2$ less than 2.10, 4.61 and 9.21, respectively.

Besides the model-independent analyses described above, extensive model-dependent theoretical studies involving the package LanHEP [61, 62] for automatic generation of Feynman rules from the Lagrangian were performed for the MSSM two-doublet Higgs sector [63]-[69].

4. Separaton of anomalous contributions to the single top production. Subsidiary fields.

In the situation when several anomalous couplings (AC) coming from different effective operators of higher dimension contribute to some particular production channel at the same time, experimental reconstruction of BSM effects in the multidimensional anomalous coupling space is technically difficult. Moreover, analyses of fancy anomalous mixtures in BSM amplitudes where the Breit-Wigner propagators are also dependent on the several AC are not meaningful with the straightforward calculation. Various anomalous contributions should be separated on the stage of unweighed events generation for the following detector simulation of the event samples which are dependent on an individual AC. Separation of congenerous contributions (e.g. of $\sim 1/\Lambda^2$ leading terms) in the events samples for experimental reconstruction is of major interest. Let us consider an example with the anomalous Wtb vertex, see Eq.(1), which includes three anomalous couplings (AC) beyond the SM. Besides the right-handed vector current coupling

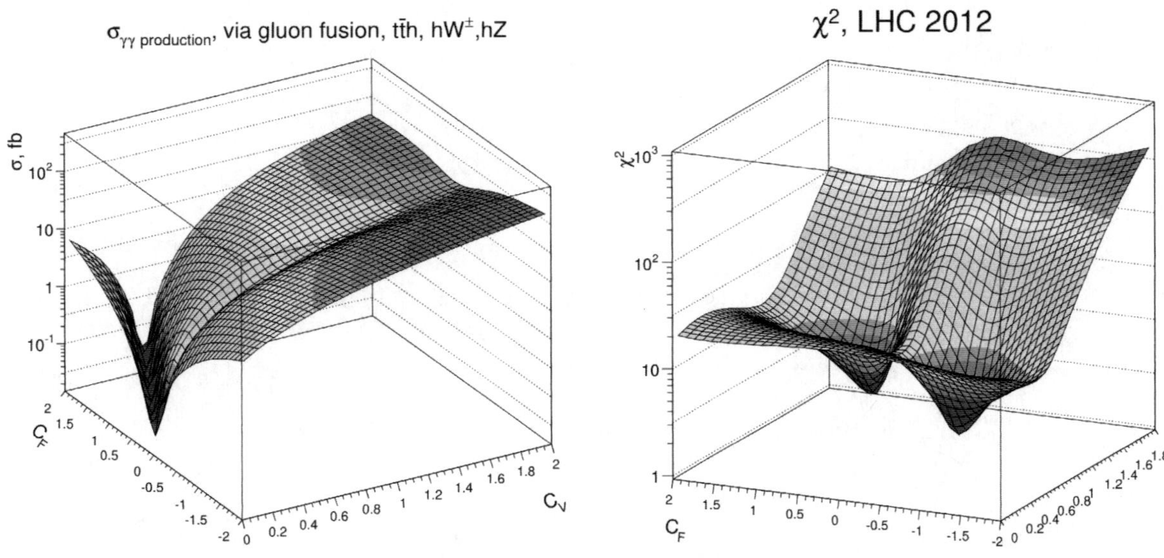

Figure 2. Three-dimensional plots in the (c_V,c_F) parameter space: (a) total cross section for $\gamma\gamma$ production processes at the LHC, $\sqrt{s}$ =8 TeV; (b) three-dimensional plot of χ^2 as a function of (c_V,c_F), LHC $\sqrt{s}$ =8 TeV - all channels

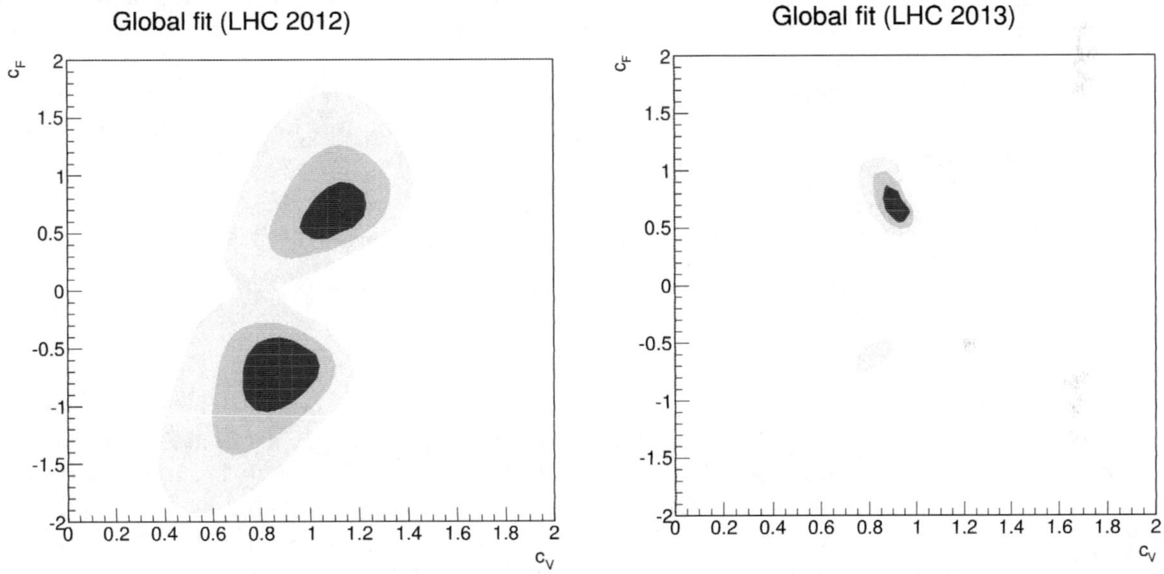

Figure 3. Left panel - three exclusion contours for the combined χ^2 fit in the (c_V,c_F) plane based on 2012 LHC data, right panel - the same based on 2013 LHC data. Blue, green and yellow areas correspond to $\Delta\chi^2$ =2.10, 4.61 and 9.21 (CL of the fit is 65%, 90% and 99%), respectively. See [50] for details.

f_{1R} it is dependent on the couplings of anomalous magnetic moment type f_{2L} and f_{2R}. So we can write

$$\Gamma_\mu \;=\; \Gamma_\mu^{SM} + \Gamma_\mu^{NP_1} + \Gamma_\mu^{NP_2} + \Gamma_\mu^{NP_3}$$

and introduce three subsidiary bosons [70] in the CompHEP Lagrangian table

W boson SM	$\frac{g}{2\sqrt{2}} f_{1L}\gamma^{\mu}(1-\gamma_5)$
W boson subsidiary 1	$\frac{g}{2\sqrt{2}} f_{1R}\gamma^{\mu}(1+\gamma_5)$
W boson subsidiary 2	$\frac{g}{2m_W\sqrt{2}} f_{2L}\sigma^{\mu\nu}q_{\nu}(1+\gamma_5)$
W boson subsidiary 3	$\frac{g}{2m_W\sqrt{2}} f_{2R}\sigma^{\mu\nu}q_{\nu}(1-\gamma_5)$

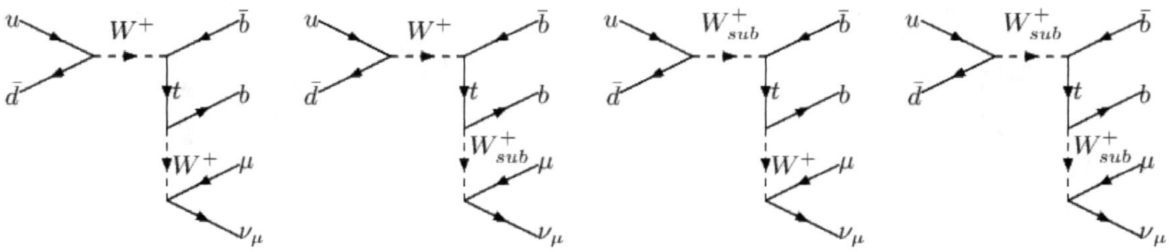

Figure 4. Diagrams (2), (3) and (4) include intermediate vector boson subsidiary fields W_{sub}.

Diagrams for the four-fermion production process $u\bar{d} \to b\bar{b}\mu^{+}\nu_{\mu}$, where anomalous Wtb vertices contribute to the production part and to the decay part of each diagram taken in the infinitely small width approximation, are shown in Fig.4. Diagrams (2),(3) and (4) include the intermediate subsidiary bosons $W_{1,2,3}$. If only f_{1R} coupling is not equal to zero in the Lagrangian Eq.(1), then the structure of matrix element squared for the sum of the four diagrams can be written in the form

$$|\mathbf{M}|^2 \sim \frac{1}{\Gamma(f_{1L},f_{1R})}[(f_{1L})^2P_1+(f_{1R})^2P_2]\times[(f_{1L})^2D_1+(f_{1R})^2D_2]$$
$$\sim \frac{1}{\Gamma(f_{1L},f_{1R})}[(f_{1L})^4P_1D_1+(f_{1L})^2(f_{1R})^2P_1D_2+(f_{1L})^2(f_{1R})^2P_2D_1+(f_{1R})^4P_2D_2]$$

where P_1 and P_2 denote the production functions dependent on the four-momenta products and D_1 and D_2 denote the decay functions. Apparently the term P_1D_2 and the term P_2D_1 are of the order of Λ^{-2} while the term P_2D_2 is of the order of Λ^{-4}. They can be easily separated after generation of all squared diagrams by the CompHEP and then omitting diagrams with the two subsidiary bosons W_{sub}. The unweighed event samples are generated for restricted sets of squared diagrams. Using a schematical denomination $(f_{1L}f_{1R}00) \leftrightarrow f_{1L}(1000)$ for an event sample where, for example, only f_{1R} anomalous coupling is taken to be non-zero, we can write

$$(f_{1L}f_{1R}00) \Leftrightarrow (f_{1L})^4\otimes(1000)\oplus(f_{1L})^2(f_{1R})^2\otimes(1100)_{sub}\oplus(f_{1R})^4\otimes(0100)^{sub}_{sub}.$$

where (0100) event sample coming from diagrams with the two subsidiary bosons does not participate in the simulation.

Data and model comparison of Baesian Neural Network discriminant (BNN) for the LHC physics analysis of the single top quark production based on the $(f_{1L}0f_{2L}0)$ event samples (left weak current SM contribution and the left anomalous tensor contribution) are shown in Fig.5. Exclusion contours in the (f_{1L}, f_{1R}) plane generated for the case $(f_{1L}f_{1R}00)$ of the right vector coupling are shown in Fig.6.

IOP Conf. Series: Journal of Physics: Conf. Series **920** (2017) 012007 doi:10.1088/1742-6596/920/1/012007

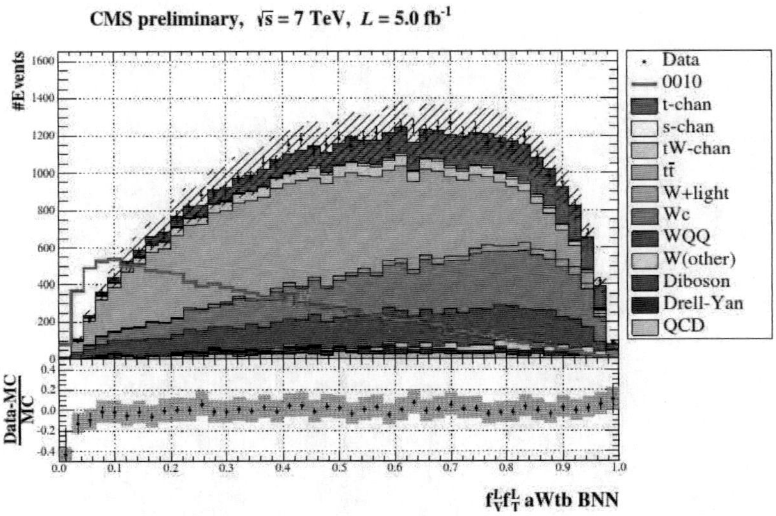

Figure 5. Data and model comparison of BNN discriminant for anomalous Wtb coupling in the case $(f_{1L}f_{1R}00)$, see more details in [71]. The BNN discriminant was trained to separate the SM events and possible events with f_{2L} coupling in the anomalous Wtb vertex. The hashed band corresponds to systematic uncertainty.

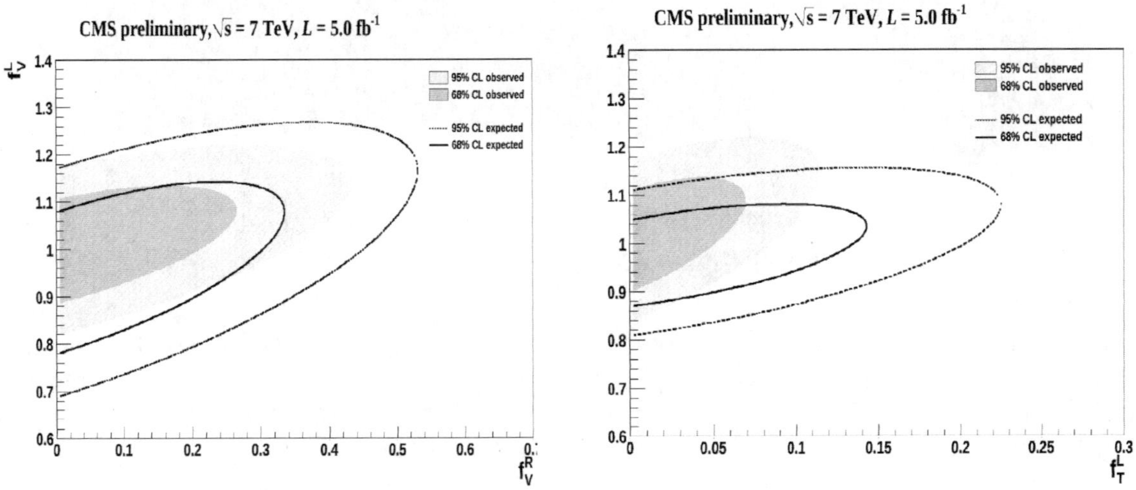

Figure 6. Exclusion contours for the top quark anomalous couplings f_{1L}, f_{1R} (left panel) and f_{2L}, f_{2R} (right panel) taken from [71].

5. New features of GUI

New features of the CompHEP Graphics User Interface (GUI) useful for generation of global fits and exclusion contours are introduced in version 4.6

- Implementation of external functions in the CompHEP Constraints Model Table
- Multiplication of selected squared diagrams on an external function
- Table calculations and algebraic operations with tables cross section/width vs parameters
- ROOT code generation to draw table functions (3D surfaces or 2D contours)
- Generation of the 3D plots for phase space distributions dependent on a model parameter

Table calculations and algebraic operations with tables are needed in the BSM analyses when the total width in the Breit-Wigner propagator of an intermediate particle depends on a number of anomalous couplings. In this case multidimensional total width table is generated in the anomalous coupling space for all partial decay channels which are then summed and used

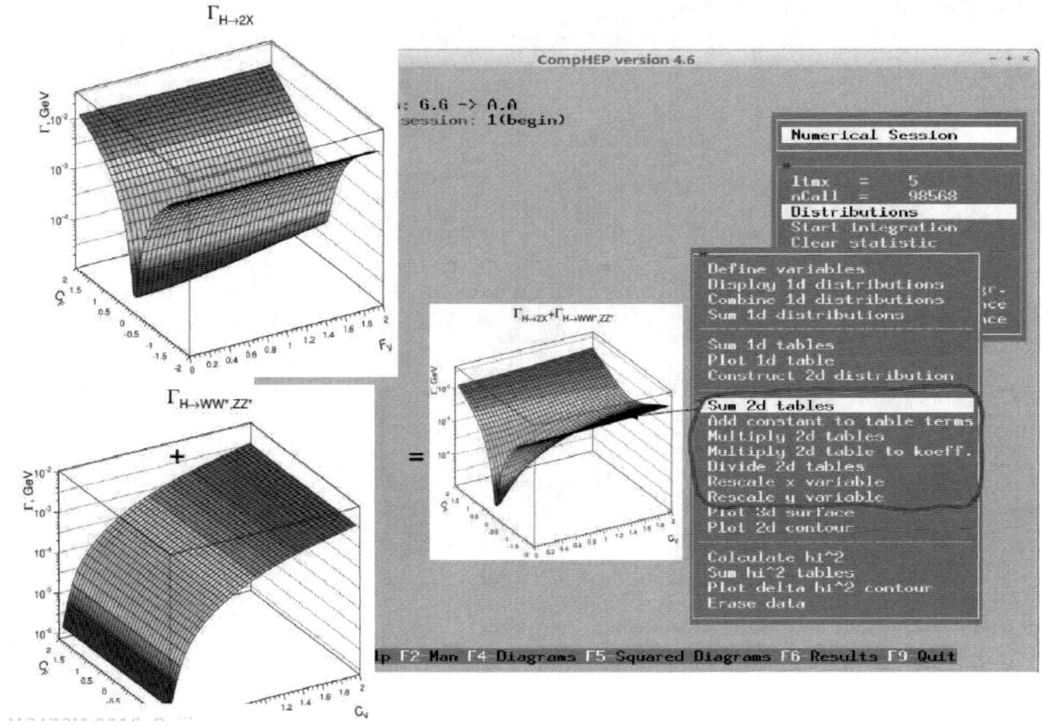

Figure 7. An example of CompHEP GUI for operation with tables for the decay width $H \to 2X$ dependent on the anomalous couplings c_V and c_F.

in the squared amplitude calculation for the signal. An example of GUI for the Higgs boson decay $H \to \gamma\gamma$ in the (c_V, c_F) coupling space (see Section 2) is depicted in Fig.7. An example of ROOT code generation for the sub-process $u\bar{d} \to t\bar{d}$ can be found in Fig.8 where the 3D t-quark distribution is generated for different values of an anomalous parameter.

6. Summary

The ideology of automatic generation of amplitudes, corresponding to the full set of Feynman rules for a given BSM Lagrangian, symbolic calculations and precise Monte Carlo integration of the complete tree-level diagram sets with subsequent generation of unweighed events demonstrated over the last two decades its efficiency for calculation of signal and background reactions at LEP2, HERA, NLC and simulation of processes for the Tevatron and the LHC.

The initial goal of the CompHEP project discussed in 1988 [1], was the establishment of an effective tool for automatic calculation of the SM signals and backgrounds at UNK (IHEP, Serpukhov) and LEP2. Although the UNK project was terminated and real practical applications started from the LEP2 physics and have been continued for linear colliders and the LHC, it is interesting to what extent the created product was the original intent. Below we reconstruct distributions of 560 citations to the NIM paper (see [33]) in the period 2004-2016, see Fig.9. Distributions of citations over a specific scientific topics and over applications by experimental collaborations are based on the INSPIRE database. Note that the package was used not in the way it had been expected. Most theoretical publications are considering extensions of the SM.

Number of studies for various extensions of the Standard Model in the CompHEP format, published by users of the package, is impressively large despite of the fact that beyond the

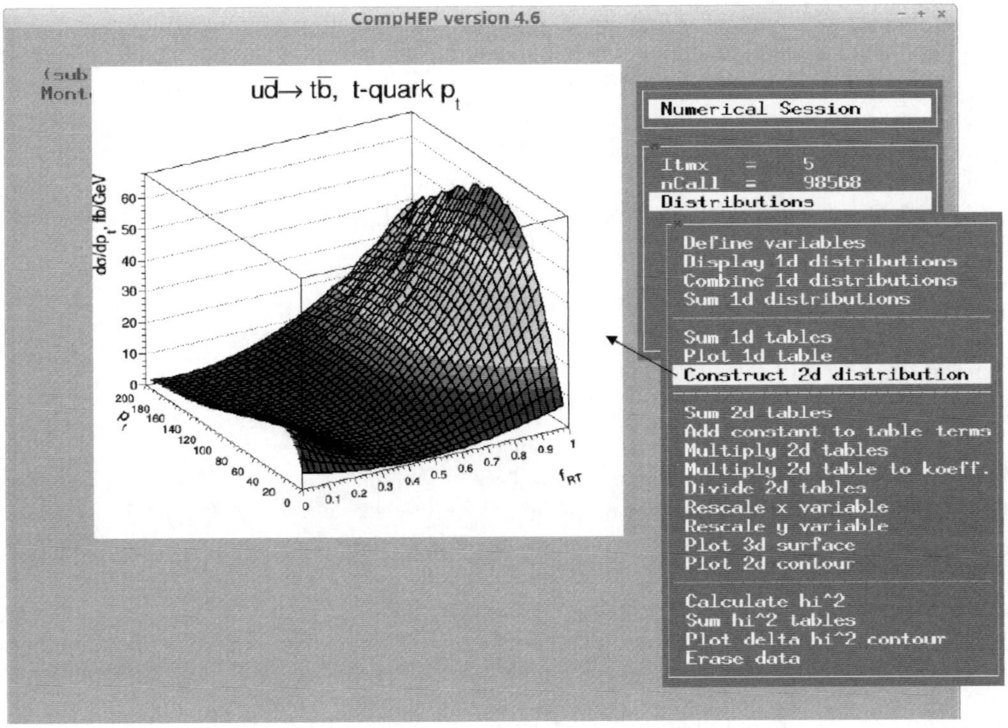

Figure 8. An example of CompHEP GUI for the ROOT code generation when the phase space distribution over p_T is dependent on an anomalous coupling.

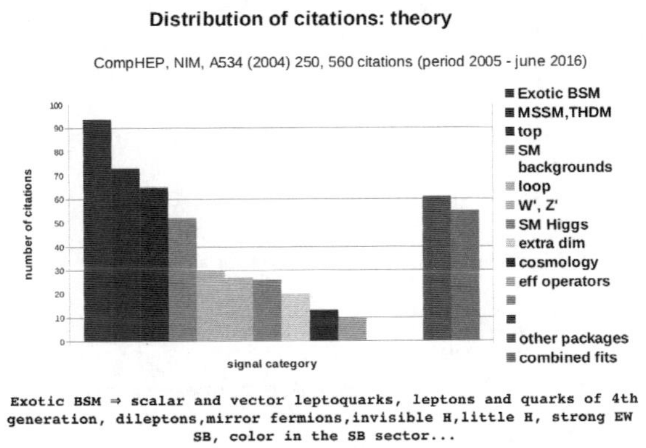

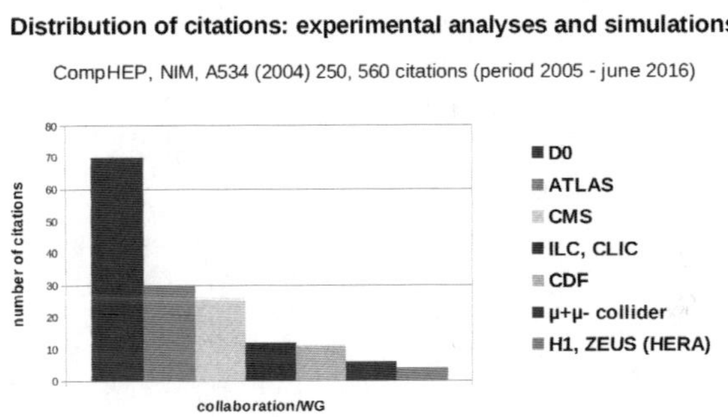

Figure 9. Distributions of theoretical and experimental citations.

'proliferation' of chiral $SU(2)$ multiplets and singlets of fundamental fermions, CompHEP version 4 provides rather restricted possibilities for work with models of higher-rank gauge symmetry groups. Rich possibilities are going to be implemented in version 5 of CompHEP, which includes the symbolic calculation kernel based on FORM [72, 73].

6.1. Acknowledgments

This proceeding is dedicated to the memory of Professor Yoshimitsu Shimizu, the coordinator of the GRACE group. His immeasurable activity in the organization of scientific cooperation between KEK and SINP MSU played a major role in the progress of work on packages of automatic calculations GRACE and CompHEP. E.B., M.D. and V.I. are grateful to the organizers of CPP 2016 for hospitality. This work was supported by Russian Science Foundation Grant No. 16-12-10280.

References

[1] Boos E E, Dubinin M N, Edneral V F, Ilyin V A, Kryukov A P, Pukhov A E, Rodionov A Ya, Savrin V I, Slavnov D A and Taranov A Yu 1989 CompHEP - computer system for calculation of particle collisions at high energies *Preprint* SINP MSU 89-63/140

[2] Boos E E, Dubinin M N, Edneral V F, Ilyin V A, Kryukov A P, Pukhov A E and Shichanin S A 1991 Applications of CompHEP system to particle processes calculation, *Preprint* SINP MSU 91-9/213

[3] Hearn A C 1993 REDUCE *user's manual, version 3.5* (RAND publication)

[4] Boos E E, Dubinin M N, Ilyin V A, Pukhov A E and Savrin V I 1994 CompHEP: Specialized package for automatic calculations of elementary particle decays and collisions INP MSU Report 94-36-358 (*Preprint* hep-ph/9503280)

[5] Kawabata S 1986 A New Monte Carlo Event Generator for High-Energy Physics *Comp.Phys.Commun.* **41** 127

[6] Kawabata S 1992 *Monte Carlo integration packages BASES and DICE, New Computing Techniques in Physics Research, Proc. of 2nd International Workshop on Software Engineering, Artificial Intelligence and Expert Systems for High-Energy and Nuclear Physics* (River Edge, N.J.: World Scientific) ed D Perret-Gallix p 745

[7] Ishikawa T, Kaneko T, Kato K, Kawabata S, Shimizu Y and Tanaka H 1993 *GRACE manual: Automatic generation of tree amplitudes in Standard Models: Version 1.0* (Tsukuba:KEK) KEK report 92-19

[8] Boos E, Dubinin M, Edneral V, Ilyin V, Pukhov A, Shichanin S, Kaneko T, Kawabata S, Kurihara Y, Shimizu Y and Tanaka H 1994 Automatic calculation in high-energy physics by Grace/Chanel and CompHEP *Int.J.Mod.Phys.* C **5** 615

[9] Boos E, Dubinin M, Ilyin V and Pukhov A 1993 e^+e^- *Collisions at 500 GeV: the Physics Potential. Proc. of the Workshop Munich, Annecy, Hamburg* ed P Zerwas (Hamburg:DESY) DESY report 93-123C, p 561

[10] Bardin D, Kleiss R et al. 1996 *Physics at LEP2*, ed G Altarelli, T Sjostrand and F Zwirner (Geneva:CERN) CERN report 96-01 vol I p 3 (*Preprint* hep-ph/9709270)

[11] Boudjema F, Mele B et al. 1996 *Physics at LEP2*, ed G Altarelli, T Sjostrand and F Zwirner (Geneva:CERN) CERN report 96-01 vol I p 207 (*Preprint* hep-ph/9601224)

[12] Carena M, Zerwas P et al, 1996 *Physics at LEP2*, ed G Altarelli, T Sjostrand and F Zwirner (Geneva:CERN) CERN report 96-01 vol II p 351 (*Preprint* hep-ph/9602250)

[13] Mangano M et al, 1996 *Physics at LEP2*, ed G Altarelli, T Sjostrand and F Zwirner (Geneva:CERN) CERN report 96-01 vol II p 299 (*Preprint* hep-ph/9602203)

[14] Grunewald M, Passarino G et al. 2000 *Reports of the Working Groups on Precision Calculations for LEP2 Physics* (Geneva:CERN) CERN report 009-A p 1 (*Preprint* hep-ph/0005309)

[15] Boos E, Dudko L and Perfilov M 2007 Interference between W' and W in single-top quark production processes *Phys. Lett.* B **655** 245 (*Preprint* hep-ph/0610080)

[16] Boos E, Dubinin M, Pukhov A, Sachwitz M and Schreiber H J 2001 Single top production in e^+e^-, e^-e^-, γe and $\gamma\gamma$ collisions. *Eur.Phys.J.* C **21** 81

[17] Boos E, Dubinin M, Pukhov A, Sachwitz M and Schreiber H J 2000 Probe of the Wtb coupling in $t\bar{t}$ pair production at linear colliders *Eur.Phys.J.* C **16** 269

[18] Boos E, Dudko L and Ohl T 1999 Complete calculations of $Wb\bar{b}$ and $Wb\bar{b}$ + jet production at Tevatron and LHC: Probing anomalous Wtb couplings in single top production *Eur.Phys.J.* C **11** 473 (*Preprint* hep-ph/9903215)

[19] Abdullin S, Dubinin M, Ilyin V, Kovalenko D, Savrin V and Stepanov N 1998 Higgs boson discovery potential of LHC in the channel $pp \to \gamma\gamma + jet$. *Phys. Lett.* B **431** 410

[20] Dubinin M N and Song H S 1998 Anomalous W boson production at HERA *Phys.Rev.* D **57** 2927

[21] Dubinin M N and Song H S 1998 W boson production at upgraded HERA *Monte Carlo generators for HERA physics* (Hamburg:DESY) p 614 (*Preprint* hep-ph/9906384)

[22] Berends F and Dubinin M 1997 Contact interactions in the four fermion processes at LEP-2 and HERA, *Proc. of 11th International Workshop on High Energy Physics and Quantum Field Theory* (Moscow:MSU) p 151 (*Preprint* hep-ph/9702344)

[23] Dubinin M 1994 Three jet events in gamma gamma collisions *Proc. of 9th International Workshop on High Energy Physics and Quantum Field Theory* (Moscow:MSU) p 135 (*Preprint* hep-ph/9504366)
[24] Dubinin M, Edneral V, Kurihara Y and Shimizu Y 1994 Complete tree level calculation of the $e^+e^- \to \nu\bar{\nu}b\bar{b}$ process and the Higgs signal at LEP200 and Next Linear Colliders *Phys. Lett.* B **329** 379
[25] Dubinin M, Kurihara Y and Shimizu Y 1994 Polarization asymmetry and the Higgs boson production mechanisms *Int.J.Mod.Phys.* A **13** 4011
[26] Boos E, Sachwitz M, Schreiber H J and Shichanin S 1994 Complete tree level calculation of the reaction $e^+e^- \to \mu^+\mu^-b\bar{b}$ and the Higgs boson signal at LEP-200 and NLC energies *Z.Phys.* C **61** 675
[27] Boos E, Dubinin M and Dudko L 1996 Higgs boson production under the resonance threshold at LEP-2 *Int.J.Mod.Phys.* A **11** 5015
[28] Dubinin M, Schreiber H J and Vologdin A 2003 Measurement of the $HZ\gamma$ coupling at the future linear e^+e^- collider *Eur.Phys.J.* C **30** 337
[29] Dubinin M 2010 Aspects of CompHEP applications to hadronic collider physics *Proc of 3rd Computational Particle Physics Workshop* (PoS:CPP 2010) 002
[30] Boos E and Dubinin M 1993 Complete tree level calculation of the $e^+e^- \to Zb\bar{b}$ process at LEP200. Higgs signal and background. *Phys. Lett.* B **308** 147
[31] Boos E and Dubinin M 2010 Problems of automatic calculation for collider physics *Phys.Usp.* **53** 1039-1051
[32] Boos E, Bunichev V, Dubinin M, Dudko L, Edneral V, Ilyin V, Kryukov A, Savrin V, Semenov A and Sherstnev A 2009 `CompHEP 4.5` *Status Report, Preprint* arXiv:0901.4757 [hep-ph])
[33] Boos E, Bunichev V, Dubinin M, Dudko L, Ilyin V, Kryukov A, Edneral V, Savrin V, Semenov A and Sherstnev A 2004 `CompHEP 4.4`: Automatic computations from Lagrangians to events *Nucl. Instrum. Meth.* A **534** 250 (*Preprint* hep-ph/0403113)
[34] Boos E, Dubinin M, Ilyin V, Pukhov A and Savrin V, *CompHEP. Specialized package for automatic calculations of elementary particle decays and collisions* (Seoul:SNU Press) SNU CTP report 94-36/358, 1994 (*Preprint* hep-ph/9503280)
[35] Pukhov A, Boos E, Dubinin M, Dudko L, Ilyin V, Kryukov A, Edneral V, Savrin V, Semenov A 1998 `CompHEP` *- a package for evaluation of Feynman diagrams and integration over multi-particle phase space. User's manual for version 3.3* (Moscow:Moscow State University Press) INP MSU report 98-41/542 (*Preprint* hep-ph/9908288)
see also `http://theory.sinp.msu.ru/comphep`
[36] Abazov V et al. (DO Collaboration) 2011 Measurements of single top quark production cross sections and $|Vtb|$ in $p\bar{p}$ collisions at $\sqrt{s}$ =1.96 TeV *Phys. Rev.* D **84** 112001
[37] Chatrchyan S et al (CMS Collaboration) 2012 Measurement of the single-top-quark t-channel cross section in pp collisions at $\sqrt{s}$ =7 TeV *JHEP* **1212** 035
[38] Khachatryan V et al. (CMS Collaboration) 2014 Measurement of the t-channel single-top-quark production cross section and of the $|Vtb|$ CKM matrix element in pp collisions at $\sqrt{s} = 8$ TeV *JHEP* **1406** 090
[39] Boos E E, Bunichev V E, Dudko L V, Savrin V I and Sherstnev A V 2006 Method for simulating electroweak top-quark production events in the NLO approximation: SingleTop event generator, *Phys. Atom. Nucl.* **69** 1317
[40] Boos E and Dudko L 2012 The single top quark physics *Int. J. Mod. Phys.* A **27** 1230026
[41] Baskakov A, Boos E, Bunichev V, Dubinin M, Dudko L, Myagkov I, Perfilov M and Vorotnikov G 2016 New Physics search in the processes of the electroweak top quark production *EPJ Web Conf* **125** 02008
[42] Chatrchyan S *et al.* (CMS Collaboration) 2012 Observation of a new boson at a mass of 125 GeV with the CMS experiment at the LHC *Phys. Lett.* B **716** 30 (*Preprint* arXiv:1207.7235[hep-ex]); 2015 *Eur. Phys. J.* C **75** 212 (*Preprint* arXiv:1412.8662[hep-ex])
[43] Aad G *et al.* (ATLAS Collaboration) 2012 Observation of a new particle in the search for the SM Higgs boson with the ATLAS detector at the LHC *Phys. Lett.* B **716** 1 (*Preprint* arXiv:1207.7214[hep-ex]); 2015 *Phys. Rev. D* **90** 052004 (*Preprint* arXiv:1406.3827[hep-ex])
[44] Gonzalez-Alonso M, Greljo A, Isidori G, Marzocca D 2015 Pseudo-observables in Higgs decays *Eur. Phys. J.* C **75** 128 (2015)
[45] Passarino G, Sturm C and Uccirati S 2010 Higgs pseudo-observables, second Riemann sheet and all that *Nucl. Phys.* B **834** 77
[46] Falkowski A and Riva F 2015 Model-independent precision constraints on dimension-6 operators *JHEP* **1502** 039
[47] Aguilar-Saavedra J A 2008 Single top quark production at LHC with anomalous Wtb couplings *Nucl.Phys.* B **804** 160
[48] Zhang C and Willenbrock S 2011 Effective-Field-Theory Approach to Top-Quark Production and Decay *Phys.Rev.* D **83** 034006
[49] Contino R, Ghezzi M, Grojean C, Muhlleitner M and Spira M Effective Lagrangian for a light Higgs-like

scalar 2013 *JHEP* **07** 035
[50] Boos E, Bunichev V, Dubinin M and Kurihara Y 2014 Higgs boson signal at complete tree level in the SM extension by dimension-six operators *Phys. Rev.* D **89** 035001
[51] Boos E, Bunichev V, Dubinin M and Kurihara Y 2014 Expectations for probing the Higgs-fermion and the Higgs- vector boson couplings at the ILC *Phys. Lett.* B **739** 410
[52] Boos E, Bunichev V and Dubinin M 2013 Combined ts of the Higgs boson signal at the LHC and the ILC by means of CompHEP, *Proc. of 21st International Workshop on High Energy Physics and Quantum Field Theory* (PoS:QFTHEP 2013) 015
[53] Buchmuller W and Wyler D 1986 Effective Lagrangian Analysis of New Interactions and Flavor Conservation *Nucl.Phys.* B **268** 621
[54] Passarino G 2013 NLO Inspired Effective Lagrangians for Higgs Physics *Nucl. Phys.* B **868** 416
[55] CMS Collaboration 2015 CERN-PH-EP/2013-037; 2015/06/16; *Preprint* arXiv:1412.8662[hep-ph]
[56] Whisnant K, Yang J M, Young B L, Zhang X 1997 *Phys. Rev.* D **56** 467 (*Preprint* hep-ph/9702305)
[57] Dumont B, Fichet S, Gersdorff G von 2013 A Bayesian view of the Higgs sector with higher dimensional operators *JHEP* **1307** 065
[58] Espinosa J R, Grojean C, Muhlleitner M and Trott M 2012 *JHEP* **1212** 045; 2012 *JHEP* **1209** 126; 2012 *JHEP* **1205** 097
[59] Kraml S, Allanach B C, Mangano M, Prosper H B, Sekmen S, Balazs C, Barr A, Bechtle P, Belanger G, Belyaev A et al. 2012 Searches for New Physics: Les Houches Recommendations for the Presentation of LHC Results *Eur.Phys.J.* C **72** 1976
[60] Azatov A, Contino R, Galloway J 2012 Model-Independent Bounds on a Light Higgs *JHEP* **1204** 127; Erratum-ibid. 2013 **1304** 140
[61] Semenov A 2016 `LanHEP` A package for automatic generation of Feynman rules from the Lagrangian. Version 3.2 *Comput.Phys.Commun.* **201** 167
[62] Semenov A 1997 `LanHEP`: A package for automatic generation of Feynman rules from the Lagrangian *Comput.Phys.Commun.* **115** 124
[63] Akhmetzyanova E, Dolgopolov M and Dubinin M 2005 Higgs bosons in the two-doublet model with CP violation *Phys.Rev.* D **71** 075008
[64] Dubinin M and Semenov A 2003 Triple and quartic interactions of Higgs bosons in the two Higgs doublet model with CP violation *Eur.Phys.J.* C **28** 223
[65] Dubinin M and Semenov A 1998 Triple and quartic interactions of Higgs bosons in the general two Higgs doublet model SNUTP-98-140; *Preprint* hep-ph/9812246
[66] Akhmetzyanova E, Dolgopolov M and Dubinin M 2006 Violation of CP invariance in the two-doublet Higgs sector of the MSSM *Phys.Part.Nucl.* **37** 677
[67] Dubinin M and Petrova E 2016 Simplified parametric scenarios of the Minimal Supersymmetric Standard Model after the discovery of the Higgs boson *Phys.Atom.Nucl.* **79** 488
[68] Dubinin M and Petrova E 2016 Post-Higgs THDM scenarios associated with the new diphoton resonance data *EPJ Web Conf* **125** 02018
[69] Dubinin M and Petrova E 2017 Radiative corrections to Higgs boson masses for the MSSM Higgs potential with dimension-six operators *Phys. Rev.* D **95** 055021
[70] Boos E, Bunichev V, Dudko L and Perfilov M 2016 Modeling of anomalous Wtb interactions in single top quark events using subsidiary fields *Int. J. Mod. Phys.* A **32** 1750008
[71] CMS Collaboration 2014 *Search for anomalous Wtb couplings and top FCNC in t-channel single-top-quark events* (CERN:Geneva) CMS Physics Analysis Summary CMS-PAS-TOP-14-007
[72] Vermaseren J A M 1991 Symbolic Manipulation with `FORM` (Amsterdam:Computer Algebra Nederlands)
[73] Ruijl B, Ueda T and Vermaseren J 2017 `FORM` version 4.2 *Preprint* arXiv:1707.06453[hep-ph]

IOP Conf. Series: Journal of Physics: Conf. Series **920** (2017) 012008 doi:10.1088/1742-6596/920/1/012008

Automated calculation of matrix elements and physics motivated observables

Z. Was

Institute of Nuclear Physics, Polish Academy of Sciences 31 342 Krakow, Poland

E-mail: z.was@cern.ch

Abstract. The central aspect of my personal scientific activity, has focused on calculations useful for interpretation of High Energy accelerator experimental results, especially in a domain of precision tests of the Standard Model. My activities started in early 80's, when computer support for algebraic manipulations was in its infancy. But already then it was important for my work. It brought a multitude of benefits, but at the price of some inconvenience for physics intuition. Calculations became more complex, work had to be distributed over teams of researchers and due to automatization, some aspects of the intermediate results became more difficult to identify.

In my talk I will not be very exhaustive, I will present examples from my personal research only: (i) calculations of spin effects for the process $e^+e^- \to \tau^+\tau^-\gamma$ at Petra/PEP energies, calculations (with the help of the Grace system of Minami-tateya group) and phenomenology of spin amplitudes for (ii) $e^+e^- \to 4f$ and for (iii) $e^+e^- \to \nu_e\bar{\nu}_e\gamma\gamma$ processes, (iv) phenomenology of CP-sensitive observables for Higgs boson parity in $H \to \tau^+\tau^-$, $\tau^\pm \to \nu 2(3)\pi$ cascade decays.

1. Introduction

Once computers equipped with algebraic languages became available, approaches to phenomenology of High Energy accelerator experiments changed substantially. For me personally, it all started in 1980. It was clear that numerous benefits appeared. Control of large expression became easier, in many cases it simply became possible for the first time. Nearly immediately some drawbacks appeared as well. For example, some seemingly obsolete expertise, like methods for special function expansions, started to disappear. At least for some years and for some communities.

This was all part of a complex and generally very fruitful development. In this presentation I will concentrate on my personal experience. I do not have any intentions to be systematic or balanced. A more balanced picture will hopefully appear together with other talks collected in the proceedings. That is also why, I think, I do not need to focus on successes of the field. These are well known. I will rather review difficulties or traps I have encountered myself. In fact traps, once resolved, turned out to be rewarding, often in an unexpected way.

The presentation is organized as follows. Section 2 will discuss computer algebra techniques which were applied in our work for Monte Carlo program KORALB [1] for the $e^+e^- \to \tau^+\tau^-\gamma$ process at Petra/PEP energies. The following Section 3, is devoted to work on spin amplitudes for KKMC Monte Carlo [2]. In particular to studies of spin amplitude sub-structure for the $e^+e^- \to \nu_e\bar{\nu}_e\gamma\gamma$ process, necessary to accommodate resulting expressions to Yennie Frautschi Suura exponentiation [3]; variant of exclusive exponentiation of initial and final state radiation

[4]. The t-channel (even though non-singular) W exchange required careful attention. The following Section 4 is devoted to phenomenology of four-fermion state production at high energy and the study of unexpected formation of peaks due to conspiracy between spin and selection cut effects [5]. Finally, in Section 5, we turn our attention to methods of Machine Learning in applications for evaluation of (massively multi-dimensional) observables for Higgs parity [6]. Section 6 closes the paper with a Summary.

2. Year 1981: KORALB Monte Carlo for $e^+e^- \to \tau^+\tau^-\gamma$ at Petra/PEP energies

At the time, Poland seemed to be an isolated place, but with enormous in-flow of young talents to research. In reality a lot of contacts existed, but it was not to be seen by me. Limited, and in fact quite awkward, access to computing existed. It looked like a hopeless loss of time, but a lot of bright minds were attracted to the computing center of Jagellonian University. I had access to algebraic manipulation language: `Shoonschip` [7] too.

One of my first projects was to evaluate the spin density matrix for the process $e^+e^- \to \tau^+\tau^-\gamma$ at Petra/PEP energies [1, 8] This work was performed under the guidance of Prof. S. Jadach. This was quite an experience in looking at spin amplitudes as (reducible) representations of (Lorentz×gauge) groups. It was a great opportunity, programs were queuing for execution time. We could then concentrate and understand the details of what we were actually doing. Later, in all my work, it came as an enormous benefit: how to represent complicated formulas (moderately complicated for today standards) of spin amplitudes in a compact form. We have used a tree of reference frames to obtain the goal, see Fig. 1. Instead of spin projections of individual fermions, we used differences (or sums) of such projections for incoming and/or outgoing leptons. In this way we could visualize spin properties of intermediate (formally virtual) photons. Simplifications of lowest order amplitudes, remained for bremsstrahlung as well. Nearly all calculations were done by hand, but every step was cross checked with the help of algebraic calculation with `Shoonschip`. This was an enormous help. We could also observe how some features of amplitudes could be identified if calculations were done by hand, but how easily these features were overlooked if we relied too much on automatization. Only if we knew what we were looking for, could we confirm the patterns with the algebraic tool. It is important to keep this in mind. Similar problems are encountered in modern times with applications of Machine Learning as well. I will return to this point in Section 5.

3. The t-channel contribution to spin amplitudes of s-channel Exclusive Exponentiation and double gluon emission in QCD.

There is a multitude of factorization schemes available for Field Theory calculations. Over years, I was checking, on particular examples, if hints for that could be identified already at the spin amplitude level. The general principle of the searches was rather simple. One tries to identify in amplitudes, gauge invariant sub-structures (parts) responsible for Matrix Element enhancements: in some regions of the phase-space (collinear-soft etc.). This is fundamental, especially from the point of view of Monte Carlo algorithm construction. Discussions with **Shimizu-sensei** were important for some stages of that work. The principle was to start from the complete expression, coded in a numerical program, to identify the most singular term and then group some other terms necessary to obtain gauge invariant part of spin amplitude. Then, for the remaining part, the search was repeated, until all interesting parts were identified. Starting point for that work had to be amplitudes guaranteed to be correct. In the two cases $e^+e^- \to \nu_e\bar{\nu}_e\gamma\gamma$ [9] and $q\bar{q} \to l^+l^-gg$ [10] expressions obtained with the help of algebraic programs were used for that purpose.

Even though in principle there were no general rules to follow, separation of amplitudes into gauge invariant parts was a straightforward and seemingly unique procedure. However I was not able to perform the task automatically. In fact, only some of the patterns appeared

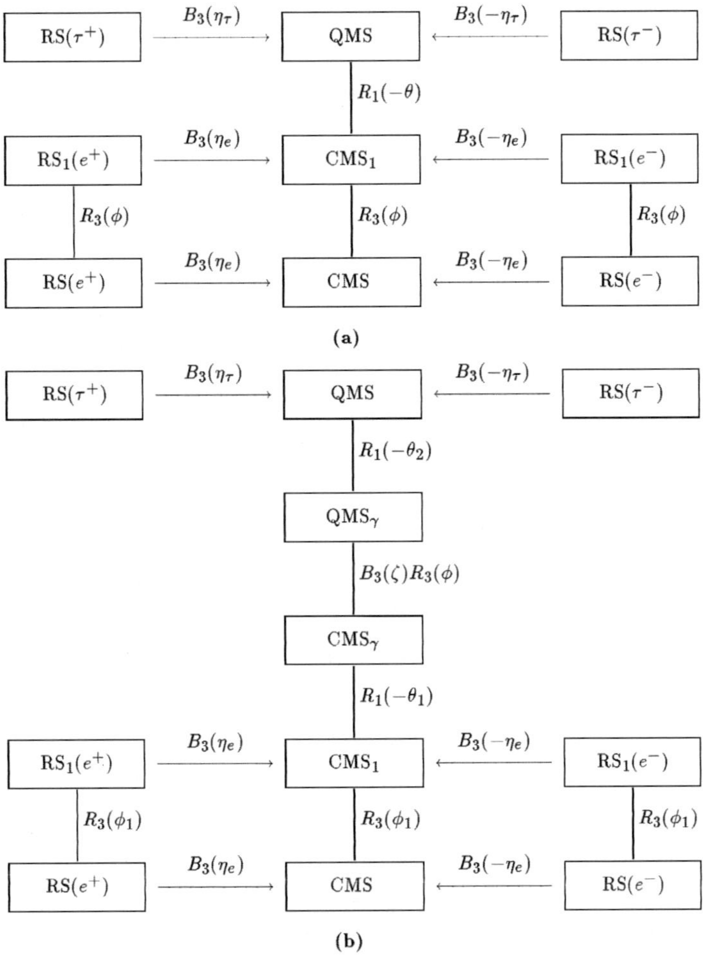

Figure 1. Tree of reference frames used to define amplitudes and Monte Carlo algorithm for $e^+e^- \to \tau^+\tau^-$ (panel a) and $e^+e^- \to \tau^+\tau^-\gamma$ (panel b) processes used and defined with all necessary details in refs. [1, 8]. Let us recall that CMS mean laboratory frame, QMS rest frame of outgoing lepton pair, RS rest frame of the indicated lepton, B_3 boost along third axis with parameter η and R_i rotation around i-axis by the given angle.

as a consequence of ordering singular terms. Feynman diagrams 1 and 2 of Fig. 2, combined, complete the amplitude for $e^+e^- \to \nu_\mu \bar{\nu}_\mu \gamma$ production, that is why they form gauge invariant part of amplitude for $e^+e^- \to \nu_e \bar{\nu}_e \gamma$ consisting of all diagrams of Fig. 2.

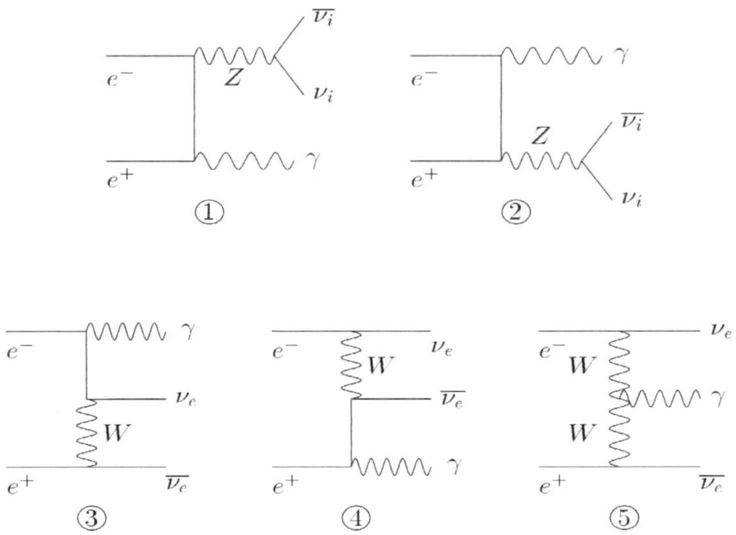

Figure 2. The Feynman diagrams for $e^+e^- \to \bar{\nu}_e \nu_e \gamma$.

Further, because these two diagrams represent initial state QED bremsstrahlung amplitude, they must divide into parts, corresponding to β_0, β_1 of Yennie-Frautshi-Suura exponentiation [3]. This has long been known, but the question was, whether it can be expanded to other cases, to higher orders or to terms of different singularities/enhancements than in case of QED photon emissions. Unexpectedly, from my experience the answer seemed to be always "yes". I could observe it not only in QED and QCD cases but also for example, in scalar QED [11]. I was encouraged by Shimizu-sensei to follow this path, also in case of complete electroweak effects, and for loop contributions, but so far I have not found the solution.

Instead, let me present first, single photon emission amplitude for $e^+e^- \to \nu_e \bar{\nu}_e \gamma$ (formula 1) and later for double gluon emission in $q\bar{q} \to l^+l^-$ (formula 2). For notation conventions references [9, 10] are used.

$$\mathcal{M}_{1\{I\}} \binom{p\, k_1}{\lambda \sigma_1} = \mathcal{M}^0 + \mathcal{M}^1 + \mathcal{M}^2 + \mathcal{M}^3, \tag{1}$$

$$\begin{aligned}
\mathcal{M}^0 &= eQ_e\, \bar{v}(p_b, \lambda_b)\, \mathbf{M}^{bd}_{\{I\}} \frac{\not{p}_a + m - \not{k}_1}{-2k_1 p_a} \not{\epsilon}^\star_{\sigma_1}(k_1)\, u(p_a, \lambda_a) \\
&\quad + eQ_e\, \bar{v}(p_b, \lambda_b)\, \not{\epsilon}^\star_{\sigma_1}(k_1) \frac{-\not{p}_b + m + \not{k}_1}{-2k_1 p_b} \mathbf{M}^{ac}_{\{I\}}\, u(p_a, \lambda_a), \\
\mathcal{M}^1 &= \mathcal{M}^{1'} + \mathcal{M}^{1''}, \\
\mathcal{M}^{1'} &= +e\, \bar{v}(p_b, \lambda_b)\, \mathbf{M}^{bd,ac}_{\{I\}}\, u(p_a, \lambda_a) \epsilon^\star_{\sigma_1}(k_1) \cdot (p_c - p_a) \frac{1}{t_a - M_W^2} \frac{1}{t_b - M_W^2}, \\
\mathcal{M}^{1''} &= +e\, \bar{v}(p_b, \lambda_b)\, \mathbf{M}^{bd,ac}_{\{I\}}\, u(p_a, \lambda_a) \epsilon^\star_{\sigma_1}(k_1) \cdot (p_b - p_d) \frac{1}{t_a - M_W^2} \frac{1}{t_b - M_W^2},
\end{aligned}$$

$$\mathcal{M}^2 = +e\,\bar{v}(p_b,\lambda_b) g^{W e\nu}_{\lambda_b,\lambda_d}\; \not\epsilon^{\star}_{\sigma_1}(k_1)\;\; v(p_d,\lambda_d)\bar{u}(p_c,\lambda_c) g^{W e\nu}_{\lambda_c,\lambda_a}\; \not k_1\;\; u(p_a,\lambda_a)\frac{1}{t_a-M_W^2}\frac{1}{t_b-M_W^2},$$
$$\mathcal{M}^3 = -e\,\bar{v}(p_b,\lambda_b) g^{W e\nu}_{\lambda_b,\lambda_d}\; \not k_1\;\; v(p_d,\lambda_d)\bar{u}(p_c,\lambda_c) g^{W e\nu}_{\lambda_c,\lambda_a}\; \not\epsilon^{\star}_{\sigma_1}(k_1)\;\; u(p_a,\lambda_a)\frac{1}{t_a-M_W^2}\frac{1}{t_b-M_W^2},$$

Once manipulations are completed, we separate the spin amplitude into *six* individually QED gauge invariant parts ($\mathcal{M}^0$ can be separated into 2 or 3 sub-parts.). This is rather easy to check, replacing photon polarization vector with its four-momentum. Each of the obtained parts has a well defined physical interpretation. It is also possible to verify that the gauge invariance of each part can be preserved in the case of the extrapolation, necessary in case of QED exclusive exponentiation (see Refs. [2, 4]). Then, because of additional photons, the condition $p_a + p_b = p_c + p_d + k_1$ is not valid.

Let us now turn to double gluon emission amplitude, that is for the process $q\bar{q} \to l^+l^-gg$:

$$\mathcal{M}^{a,b} = \frac{1}{2}\,\bar{v}(p)\Big(T^aT^bI^{(1,2)} + T^bT^aI^{(2,1)}\Big)u(q)\;. \tag{2}$$

For the T^aT^b-part, we find

$$\begin{aligned}
I^{(1,2)} &= \left(\frac{p\cdot e_1}{p\cdot k_1} - \frac{k_2\cdot e_1}{k_2\cdot k_1} - \frac{\not e_1\not k_1}{2p\cdot k_1}\right)\not J\left(\frac{\not k_2\not e_2}{2q\cdot k_2} + \frac{k_1\cdot e_2}{k_1\cdot k_2} - \frac{q\cdot e_2}{q\cdot k_2}\right)\\
&+ \frac{p\cdot k_2}{p\cdot k_1 + p\cdot k_2 - k_1\cdot k_2}\left(\frac{p\cdot e_1}{p\cdot k_1} - \frac{k_2\cdot e_1}{k_2\cdot k_1} - \frac{\not e_1\not k_1}{2p\cdot k_1}\right)\left(\frac{p\cdot e_2}{p\cdot k_2} - \frac{k_1\cdot e_2}{k_1\cdot k_2} - \frac{\not e_2\not k_2}{2p\cdot k_2}\right)\not J\\
&+ \not J\frac{q\cdot k_1}{q\cdot k_1 + q\cdot k_2 - k_1\cdot k_2}\left(\frac{q\cdot e_1}{q\cdot k_1} - \frac{k_2\cdot e_1}{k_2\cdot k_1} - \frac{\not k_1\not e_1}{2q\cdot k_1}\right)\left(\frac{q\cdot e_2}{q\cdot k_2} - \frac{k_1\cdot e_2}{k_1\cdot k_2} - \frac{\not k_2\not e_2}{2q\cdot k_2}\right)\\
&+ \not J\left(1 - \frac{p\cdot k_2}{p\cdot k_1 + p\cdot k_2 - k_1\cdot k_2} - \frac{q\cdot k_1}{q\cdot k_1 + q\cdot k_2 - k_1\cdot k_2}\right)\left(\frac{k_1\cdot e_2}{k_1\cdot k_2}\frac{k_2\cdot e_1}{k_1\cdot k_2} - \frac{e_1\cdot e_2}{k_1\cdot k_2}\right)\\
&- \frac{1}{4}\frac{1}{p\cdot k_1 + p\cdot k_2 - k_1\cdot k_2}\left(\frac{\not e_1\not k_1\not e_2\not k_2 - \not e_2\not k_2\not e_1\not k_1}{k_1\cdot k_2}\right)\not J\\
&- \frac{1}{4}\not J\frac{1}{q\cdot k_1 + q\cdot k_2 - k_1\cdot k_2}\left(\frac{\not k_1\not e_1\not k_2\not e_2 - \not k_2\not e_2\not k_1\not e_1}{k_1\cdot k_2}\right)\;.
\end{aligned} \tag{3}$$

The part proportional to other order of SU(3) group generators T^bT^a, is obtained by a permutation of the momenta and polarization vectors of the gluons. Each line of the above expression is individually gauge invariant. Also, this expression for double gluon emission is rather compact too.

4. The $e^+e^- \to 4f$ process

The main purpose of my visit to KEK MinamiTateya group in 1996, was to work on `Grace` spin amplitudes [5] and to prepare them for use in our `KORALW` Monte Carlo [12] for the $e^+e^- \to 4$ fermion processes at LEP II energies. In this work, because of Monte Carlo integration, phase space regions of collinear configurations, resulted in numerical difficulties. This required careful and painful work to avoid 'trivial' mistakes due to rounding errors damaging gauge cancellations. On the margins of this work, kinds of fake 'New Physics', phenomena appeared. Let me show, for myself at first rather unexpected example, see Ref. [13]. It illustrates that the interplay of theoretical effects and selection cuts can be confusing. In $e^+e^- \to WW \to q\bar{q}q\bar{q}$ two jets were requested to be lost in the beam pipe the invariant mass of the other two was monitored. If all diagrams of the lowest order Standard Model amplitudes were taken into account, a clear Z peak is present (Fig. 3 left side). But there is another peak at a much higher mass, which do not disappear, even if only W-pair production and decay amplitudes are taken

IOP Conf. Series: Journal of Physics: Conf. Series **920** (2017) 012008 doi:10.1088/1742-6596/920/1/012008

The $\frac{d\sigma_2}{dM_{s\bar{s}}}$ differential distribution of the "visible" $s\bar{s}$ jets where $c\bar{c}$ jets escape detection. The centre-of-mass energy is 195 GeV. Input parameters of type 2: CC-03 (thick line); and type 4: CC-43 (thin line).

The $\frac{d\sigma_2}{dM_{s\bar{s}}}$ differential distribution of the "visible" $s\bar{s}$ jets where $c\bar{c}$ jets escape detection. The centre-of-mass energy is 195 GeV. Input parameters of type 1: CC-03 no spin correlation (thin line); and type 2: CC-03 spin correlations switched on (thick line).

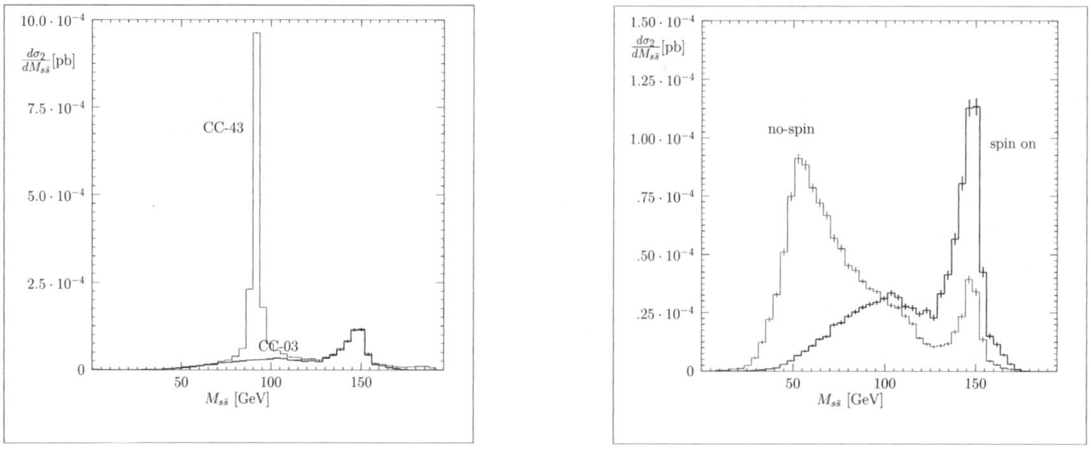

Figure 3. Figure taken from Ref. [13]

into account (see Fig. 3 right side). It is a consequence of the veto cut, W^+ and W^- peaks, and spin correlations as explained in Ref. [13].

Such phenomena need to be included in generators used for simulation of backgrounds. Otherwise intuition may fail to recognize importance for background estimation effects such as generally small spin correlations. Cross-check with `Grace` spin amplitudes was helpful to confirm the origin of the phenomenon.

5. Machine Learning for Higgs parity measurement in $H \to \tau\tau$ decay

Let us now turn to another example, where complex observables need to be defined and computer techniques of the Neural Network or Machine Learning (ML) type [14, 15, 16, 17, 18] are useful or even irreplaceable. Fifteen years ago, in Ref. [19], we have proposed to measure Higgs boson parity in its $H \to \tau\tau$ decay with the help of acoplanarity angle for planes built on visible decay products for consecutive $\tau^\pm \to \pi^\pm\pi^0\nu_\tau$ decays. A variant of the observable where impact parameter was used, is presented in [20]. In attempt to extend the method to $\tau \to 3\pi\nu$ decays, one can construct 4 or 16 such angles for each event. Each of these distributions provide some CP sensitivity, but distributions are correlated and possible backgrounds will complicate future measurements even further. Such observables may be controlled today with ML techniques. All correlations between angles and y_i variables (defined later) can be automatically identified and explored in an optimal way. The task nearly impossible to perform by hand. An attempt in this direction was presented in Ref. [6]. In the following, let us recall some details of this work.

The H or A (scalar or pseudo-scalar), thanks to the sign difference in the formula below, can be extracted from the correlations between τ^+ and τ^- transverse spin components. The decay probability

$$\Gamma(H/A \to \tau^+\tau^-) \sim 1 - s_\parallel^{\tau^+} s_\parallel^{\tau^-} \pm s_\perp^{\tau^+} s_\perp^{\tau^-}$$

is sensitive to the $\tau^\pm$ polarization vectors s^{τ^-} and s^{τ^+} (defined in their respective rest frames). The symbols $\parallel,\perp$ denote components parallel/transverse to the Higgs boson momentum as seen from the respective $\tau^\pm$ rest frames.

Because of the narrow τ width, cross-section for the process $f\bar{f} \to \tau^+\tau^- Y; \tau^+ \to X^+\bar{\nu}; \tau^- \to X^-\nu$ reads:

$$d\sigma = \sum_{spin} |\mathcal{M}|^2 d\Omega = \sum_{spin} |\mathcal{M}|^2 d\Omega_{prod}\, d\Omega_{\tau^+}\, d\Omega_{\tau^-}.$$

With only τ spin indices explicit $\mathcal{M}$ reads:

$$\mathcal{M} = \sum_{\lambda_1\lambda_2=1}^{2} \mathcal{M}^{prod}_{\lambda_1\lambda_2}\, \mathcal{M}^{\tau^+}_{\lambda_1}\, \mathcal{M}^{\tau^-}_{\lambda_2}.$$

The $d\sigma$ can be re-written into core formula of spin algorithms of τ-pair production and decay:

$$d\sigma = \Big(\sum_{spin} |\mathcal{M}^{prod}|^2\Big)\Big(\sum_{spin} |\mathcal{M}^{\tau^+}|^2\Big)\Big(\sum_{spin} |\mathcal{M}^{\tau^-}|^2\Big) wt\; d\Omega_{prod}\, d\Omega_{\tau^+}\, d\Omega_{\tau^-}$$

To complete explanations, we need to first recall, details of general formalism of semileptonic τ-lepton decays. The Matrix Element used in TAUOLA Monte Carlo [21] for semileptonic decay $\tau(P,s) \to \nu_\tau(N)X$ can be written as follows: $\mathcal{M} = \frac{G}{\sqrt{2}}\bar{u}(N)\gamma^\mu(v + a\gamma_5)u(P)J_\mu$, the current J_μ depends on the momenta of all hadrons. Then

$$\begin{aligned}
|\mathcal{M}|^2 &= G^2\frac{v^2+a^2}{2}(\omega + H_\mu s^\mu), \qquad (4)\\
\omega &= P^\mu(\Pi_\mu - \gamma_{va}\Pi^5_\mu),\\
H_\mu &= \frac{1}{M}(M^2\delta^\nu_\mu - P_\mu P^\nu)(\Pi^5_\nu - \gamma_{va}\Pi_\nu),\\
\Pi_\mu &= 2[(J^*\cdot N)J_\mu + (J\cdot N)J^*_\mu - (J^*\cdot J)N_\mu],\\
\Pi^{5\mu} &= 2\ \mathrm{Im}\ \epsilon^{\mu\nu\rho\sigma}J^*_\nu J_\rho N_\sigma,\\
\gamma_{va} &= -\frac{2va}{v^2+a^2},\\
\hat{\omega} &= 2\frac{v^2-a^2}{v^2+a^2}m_\nu M(J^*\cdot J),\\
\hat{H}^\mu &= -2\frac{v^2-a^2}{v^2+a^2}m_\nu\ \mathrm{Im}\ \epsilon^{\mu\nu\rho\sigma}J^*_\nu J_\rho P_\sigma.
\end{aligned}$$

In the following we will use $h_i = H^i/H_0$.

The Higgs decay probability in the formalism of Refs. [22, 23] in case when both scalar and pseudo-scalar $H\tau\tau$ coupling are allowed $\bar{\tau}N(\cos\phi^{CP} + i\sin\phi^{CP}\gamma_5)\tau$, reads

$$\Gamma(h_{mix} \to \tau^+\tau^-) \sim 1 - s^{\tau^+}_{\parallel} s^{\tau^-}_{\parallel} + s^{\tau^+}_{\perp} R(2\phi^{CP})\ s^{\tau^-}_{\perp}. \qquad (5)$$

The $R(2\phi^{CP})$ – denotes the operator for the rotation by angle $2\phi^{CP}$ around the $\parallel$ direction.

As a consequence, spin weight wt for the combined production and decay of tau pair reads as:

$$wt = 1 - h^{\tau^+}_{\parallel} h^{\tau^-}_{\parallel} + h^{\tau^+}_{\perp} R(2\phi^{CP})\ h^{\tau^-}_{\perp}. \qquad (6)$$

Naturally, the Higgs parity should reflects itself in some kind of correlations between the τ^+ and τ^- decay products in the directions transverse to the $\tau^+\tau^-$ axes.

Let us recall first the case when both τ leptons decay to $\pi^\pm\pi^0\nu_\tau$. Then $h^i = \mathcal{N}\Big(2(q\cdot N)q^i - q^2N^i\Big)$, where $q\cdot N = (E_{\pi^\pm} - E_{\pi^0})m_\tau$ and four momentum $q = p_{\pi^\pm} - p_{\pi^0}$ is build from

IOP Conf. Series: Journal of Physics: Conf. Series **920** (2017) 012008 doi:10.1088/1742-6596/920/1/012008

four momenta of π's. The N denote four momentum of neutrino again in the τ lepton rest-frame. Because $q \cdot N$ may be either positive or negative, corresponding regions of phase-space would contribute canceling out parity effects. To separate we may use y_1, y_2 variables. The $y_1 = \frac{E_{\pi^+} - E_{\pi^0}}{E_{\pi^+} + E_{\pi^0}}$; $y_2 = \frac{E_{\pi^-} - E_{\pi^0}}{E_{\pi^-} + E_{\pi^0}}$ can be calculated directly from π's energies as measured in the laboratory frame. It is enough to take the sign of the $y_1 y_2$ only. The sensitivity to parity can be seen in Fig. 4. Acoplanarity angle (see Fig. 5 for definition) was used. Already in this case, the observable is in principle of 3-dimensional nature.

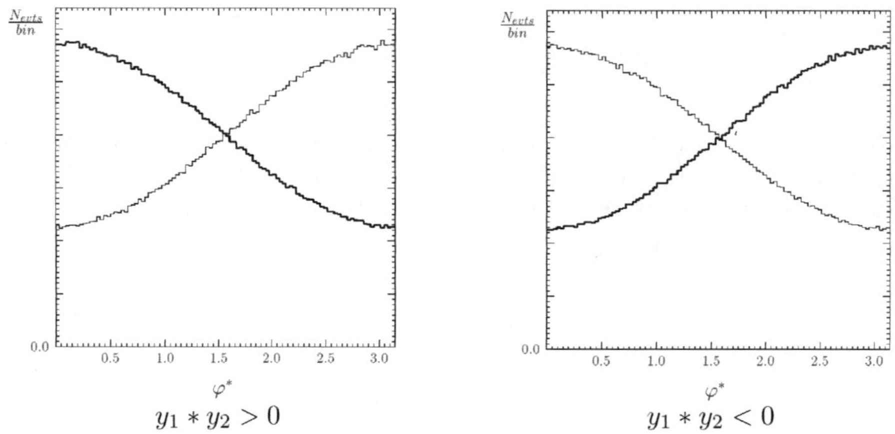

Figure 4. The $\rho^+\rho^-$ decay products' acoplanarity distribution in $\rho^+\rho^-$ pair rest-frame. Thick line denote the case of the scalar Higgs and thin line the pseudo-scalar.

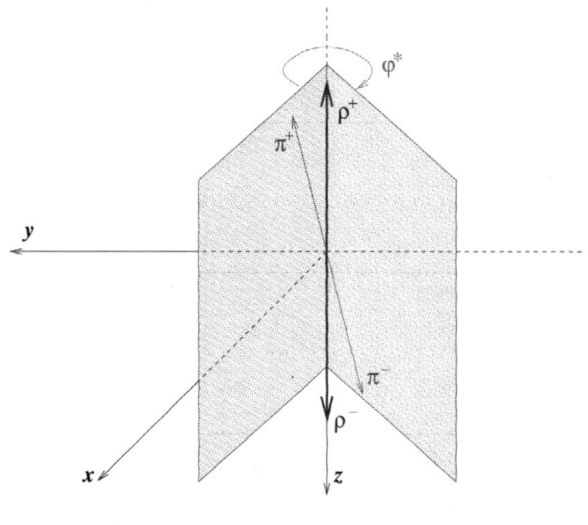

Figure 5. Acoplanarity angle $0 < \varphi^* < 2\pi$ between oriented half-planes (or $0 < \varphi^* < \pi$ between oriented planes) spanned respectively on $\rho^\pm - \pi^\pm$ decay products and in the rest frame of $\rho^+\rho^-$ pair.

In case of $\tau \to 3\pi\nu$ decay products, *four* distinct planes can be spanned on its visible decay products, thus if both τ^+ and τ^- decay in this Chanel, 16 acoplanarity angles can be defined, see Fig. 6, and also 8 variables similar to $y_{1,2}$. The observable is thus built on 24 dimensional space. Fortunately, with the modern techniques, such as used in Ref. [6] an overall sensitivity can be evaluated. On the other hand, as it may be difficult to develop intuition, the risk of misinterpretation can not be ignored. Difficulties of pattern recognition projects are known, see Fig. 7. These challenges, were explored by Giuseppe Arcimboldo (1572 - 1593) in his paintings.

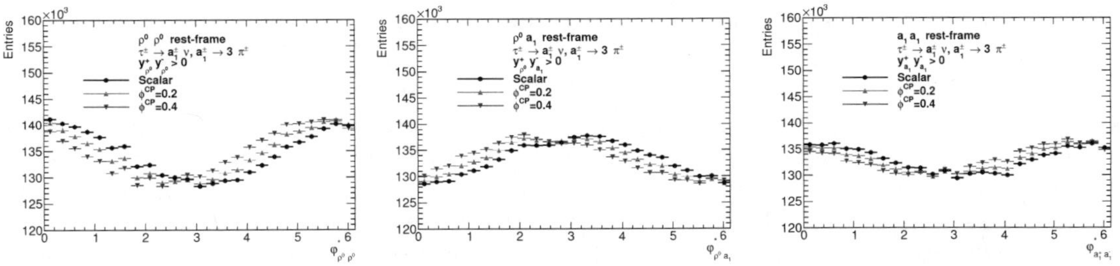

Figure 6. Acoplanarity angles of oriented half decay planes: $\varphi^*_{\rho^0\rho^0}$ (left), $\varphi^*_{a_1\rho^0}$ (middle) and $\varphi^*_{a_1a_1}$ (right), for events grouped by the sign of $y^+_{\rho^0}y^-_{\rho^0}$, $y^+_{a_1}y^-_{\rho^0}$ and $y^+_{a_1}y^-_{a_1}$ respectively. Three CP mixing angles $\phi^{CP} = 0.0$ (scalar), 0.2 and 0.4. Note that physics model depends on 1 parameter only and effect of ϕ^{CP}, the Higgs mixing scalar pseudo-scalar angle, is a linear shift.

Figure 7. Artificial Neural Networks have spurred remarkable recent progress in image classification and speech recognition. But even though these are very useful tools based on well-known mathematical methods, we actually understand surprisingly little of why certain models work and others don't.
From http://googleresearch.blogspot.com/2015/06/inceptionism-going-deeper-into-neural.html
Pattern recognition is an active field and deep concern and not only for us.

6. Summary

Inspired by Shimizu-sensei conference, I have focused on an aspect of work for Monte Carlo generators and on phenomenology of High Energy Physics experiments related to the use of

IOP Conf. Series: Journal of Physics: Conf. Series **920** (2017) 012008 doi:10.1088/1742-6596/920/1/012008

algebraic manipulation programs. My aim was to show several simple examples of challenges, resulting from complexity: how automated calculations were of help, but also sources of difficulties. Some of the important examples originate from my work in Minami Tateya group I was visiting over the last 25 years. Each example deserve substantial introduction. This was impossible for a short talk. Whenever possible, I delegate the reader to references. Inescapably, I have presented only scattered projects, where use of computer algebraic methods or pattern recognition techniques (Machine Learning) were necessary. Collecting material for my talk was inspiring and educative for myself as I could look back at particular context of my scientific activities over all these years.

Acknowledgments

This talk would not be possible without contribution and collaboration with the co-authors of the presented papers: S. Jadach, E. Richter-Was, M. Skrzypek and B. F. L. Ward. I would like to stress importance of research which was performed in collaboration with T. Ishikawa and Y. Kurihara at times when I was visiting Minanmi Tateya group of Prof. Y. Shimizu. This collaboration was essential part of my experience with algebraic manipulation programs and programs constructed with their help.

Work supported in part by funds of Polish National Science Centre under decisions UMO-2014/15/ST2/00049 and by the Research Executive Agency (REA) of the European Union under the Grant Agreement PITNGA2012316704 (HiggsTools). I would like to express gratitude to the conference organizers for the invitation.

References

[1] Jadach S and Was Z 1985 *Comput. Phys. Commun.* **36** 191–211
[2] Jadach S, Ward B F L and Was Z 2000 *Comput. Phys. Commun.* **130** 260–325 (*Preprint* hep-ph/9912214)
[3] Yennie D R, Frautschi S and Suura H 1961 *Ann. Phys. (NY)* **13** 379
[4] Jadach S, Ward B F L and Was Z 2001 *Phys. Rev.* **D63** 113009 (*Preprint* hep-ph/0006359)
[5] Fujimoto J *et al.* 2003 *Comput. Phys. Commun.* **153** 106–134 (*Preprint* hep-ph/0208036)
[6] Jzefowicz R, Richter-Was E and Was Z 2016 *Phys. Rev.* **D94** 093001 (*Preprint* 1608.02609)
[7] Veltman M 1965, an IBM Programme for Symbolic Manipulation of Algebriac Expressions, Especially Feynman Diagrams (CERN, unpublished).
[8] Jadach S and Was Z 1984 *Acta Phys. Polon.* **B15** 1151 [Erratum: Acta Phys. Polon.B16,483(1985)]
[9] Was Z 2005 *Eur. Phys. J.* **C44** 489–503 (*Preprint* hep-ph/0406045)
[10] van Hameren A and Was Z 2009 *Eur. Phys. J.* **C61** 33–49 (*Preprint* 0802.2182)
[11] Nanava G and Was Z 2007 *Eur. Phys. J.* **C51** 569–583 (*Preprint* hep-ph/0607019)
[12] Jadach S, Placzek W, Skrzypek M, Ward B F L and Was Z 2001 *Comput. Phys. Commun.* **140** 475–512 (*Preprint* hep-ph/0104049)
[13] Ishikawa T, Kurihara Y, Skrzypek M and Was Z 1998 *Eur. Phys. J.* **C4** 75–84 (*Preprint* hep-ph/9702249)
[14] LeCun Y, Bengio Y and Hinton G 2015 *Nature* **521** 436–444
[15] Kingma D and Ba J 2014 *arXiv preprint arXiv:1412.6980*
[16] Ioffe S and Szegedy C 2015 *arXiv preprint arXiv:1502.03167*
[17] Srivastava N, Hinton G E, Krizhevsky A, Sutskever I and Salakhutdinov R 2014 *Journal of Machine Learning Research* **15** 1929–1958
[18] Abadi M, Agarwal A, Barham P, Brevdo E, Chen Z, Citro C, Corrado G S, Davis A, Dean J, Devin M *et al.* 2015 *Software available from tensorflow. org* **1**
[19] Bower G R, Pierzchala T, Was Z and Worek M 2002 *Phys. Lett.* **B543** 227–234 (*Preprint* hep-ph/0204292)
[20] Desch K, Was Z and Worek M 2003 *Eur. Phys. J.* **C29** 491–496 (*Preprint* hep-ph/0302046)
[21] Jadach S, Was Z, Decker R and Kuhn J H 1993 *Comput. Phys. Commun.* **76** 361–380
[22] Kramer M, Kuhn J H, Stong M L and Zerwas P M 1994 *Z. Phys.* **C64** 21–30 (*Preprint* hep-ph/9404280)
[23] Desch K, Imhof A, Was Z and Worek M 2004 *Phys. Lett.* **B579** 157–164 (*Preprint* hep-ph/0307331)

IOP Conf. Series: Journal of Physics: Conf. Series **920** (2017) 012009 doi:10.1088/1742-6596/920/1/012009

Automatic calculation of two-loop ELWK corrections to the muon (g-2)

Tadashi Ishikawa[1], **Nobuya Nakazawa**[2] **and Yoshiaki Yasui**[3]

[1] High Energy Accelerator Organization(KEK), 1-1 OHO Tsukuba Ibaraki 305-0801,Japan
[2] Department of Physics, Kogakuin University, Shinjuku,Tokyo 163-8677,Japan
[3] Department of Management, Tokyo Management College,Ichikawa, Chiba 272-0001, Japan

E-mail: nobuya@suchix.kek.jp

Abstract. Two-loop electroweak corrections to the muon anomalous magnetic moment are automatically calculated by using GRACE-FORM system, as a trial to extend our system for two-loop calculation. We adopt the non-linear gauge (NLG) to check the reliability of our calculation. In total 1780 two-loop diagrams consisting of 14 different topological types and 70 one-loop diagrams composed of counter terms are calculated. We check UV- and IR-divergences cancellation and the independence of the results from NLG parameters. As for the numerical calculation, we adopt trapezoidal rule with Double Exponential method (DE). Linear extrapolation method (LE) is introduced to regularize UV- and IR- divergences and to get finite values.

1. Introduction

In order to get a sign of beyond the standard model physics from high precision experimental data, we need higher order radiative corrections within Standard Model (SM). For this purpose our group has been developing the automatic calculation system GRACE [1] since the late 1980's. The measurement of the muon anomalous magnetic moment (g-2) is the one of the most precise experiments to check the SM. QED correction was calculated by T.Kinoshita et al. [2] up to tenth-order. The two-loop electroweak (ELWK) correction to (g-2) was calculated approximately by Kukhto et al. [3] in 1992. Surprisingly, the two-loop correction is almost 20% of the one-loop correction. We started to calculate the full two-loop corrections in 1995 and presented our formalism at Pisa conference [4]. We also showed that the two-loop QED value [5] was correctly reproduced within our general formalism. However, the number of diagrams is huge (1780+70) and the numerical integration requires the big CPU-power to achieve required accuracy, we must wait until various environments are improved.

During these days, the several groups did the calculations using leading $\log(M^2)$ approximations. (M=heavy particle mass) [6] [7] and the approximate value of the two-loop ELWK correction is widely accepted [8]. In 2001, BNL-Experiment 821 [9] announced that the precise experimental value deviates from that of SM around $(2.2 \sim 2.7)\ \sigma$. It brought much interest in the theoretical value. The various theoretical predictions and experimental value are summarized in the next subsection. The main theoretical concern is shifted to the hadronic contributions.

1.1. present status of the theoretical predictions of (g-2) and future experiments
We summarize the present theoretical predictions of muon (g-2) in Table 1.

Table 1. Theoretical predictions of muon (g-2) [10]

Type of correction	Numerical Value (unit 10^{-11})	Error	Reference
QED up to tenth-order	116584718.95	(0.08)	[2]
Leading Order Hadronic Vac.Pol.	6923	(42)(3)	[11]
NLO Vac.Pol.+Hadronic LBL	7	(26)	[12]
ELWK (one-loop)	195.82	(0.02)	[13]
ELWK (two-loop)	- 41.2	(1.0)	[3][6][7][14]
Theory total	116591803	(1)(42)(26)	[10]
Experimental Value	116592091	(54)(33)	[9][15]

The discrepancy between the experimental value and the theoretical value is still large. As new experiments are scheduled at FNAL-E989 [16] and J-PARC-E034 [17], we can expect to have new data within a few years.

1.2. Purpose of our calculation
Although the two-loop ELWK correction is almost established as summarized in the Table 1, we try to get the value without any approximation to confirm the validity of the earlier studies. It is also an important trial to extend GRACE-system from one-loop to two-loop calculation. It may become a milestone to construct the framework of Perturbative Numerical Quantum Field Theory (PNQFT). The concepts of PNQFT are summarized as follows. (a) It is essential to assume amplitudes as meromorphic functions of space time dimension n, for extracting both UV- and IR-divergences. By adopting dimensional regularization, gauge invariant renormalized value is obtained directly. (b) The source for numerical integration is automatically generated by a symbolic manipulation system. (c) It is crucial to reduce human intervention to avoid careless mistakes, so that numerical method is fully exploited. The following calculation shows the outline of this direction.

In section 2 and 3, we briefly explain the flow and foundation of our calculation. In section 4, we touch on our method of numerical calculation. We emphasize that the Linear Extrapolation (LE) method [18] is simple and efficient method to regularize UV- and IR-divergences and also to get finite values. In numerical calculation, the validity of the results must be guaranteed by comparing several independent methods. We explain shortly our consistency conditions to ensure the results. We also show some examples of calculations and how the consistency conditions are satisfied. In section 5, we give a part of our results on (g-2) for restricted types of diagrams. In the last section, we summarize the present status and give some comments to make extensive progress.

2. Outline of our frame work

Our calculation is formulated under the following conditions.

(i) Adopt non-linear gauge (NLG) formulation with 't Hooft-Feynman propagator.
(ii) Dimensional regularization for both Ultra Violet (UV)- and Infrared (IR) -divergences.
(iii) On mass shell renormalization scheme is adopted.

(iv) Linear Extrapolation method (LE) is fully used for regularization and getting finite values.

Next, we briefly explain the flow of our calculation.

(i) GRACE system generates all the diagrams we need in SM. There are 1780 two-loop diagrams and 70 one-loop diagrams composed of one-loop order counter term (CT).
(ii) These 1780 diagrams are classified into 14 types of topology. Types of the topology are displayed in Fig.1. Among these types, some of them give the same contribution because of symmetry. (an example: 5-a vs. 5-b) The diagrams including CT are essentially two types, namely, vertex and self-energy types.
(iii) In order to distinguish the UV-part from finite part, we prepare a few files determined by the type of topology, in advance. They represent the following quantities. [19]

- Internal loop momentum flow($\ell_s, \eta(s)$) ($s = 1, 2$)
- External momentum flow (q_j) (j = internal line number)
- Kirchhoff's law of momentum conservation at each vertex

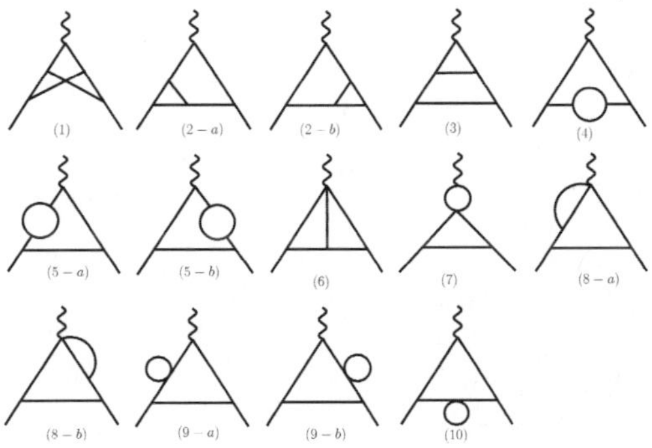

Figure 1. Types of topology

(iv) Using these tools, both UV-divergent and finite contributions to (g-2) are calculated in turn, according to the formulas given in the next section [19]. We make use of a symbolic manipulation system FORM [20] exhaustively.

3. Logic of the calculation

We explain the case where there are six-propagators. Starting formula is the two-loop muon vertex,

$$\Gamma_\mu = \int \frac{d^n\ell_1}{i(2\pi)^n}\frac{d^n\ell_2}{i(2\pi)^n}\frac{F_\mu(D)}{\prod_j(p_j^2 - m_j^2)} = \Gamma(6)\int\prod dz_j\delta(1-\sum_j z_j)\int\frac{d^n\ell_1}{i(2\pi)^n}\frac{d^n\ell_2}{i(2\pi)^n}\frac{F_\mu(D)}{\sum_j z_j(p_j^2 - m_j^2)} \quad (1)$$

where $p_j = \sum_{s=1}^{2}\eta(s)\ell_s + q_j$, is the momentum on the internal line (j). The function $\eta(s)(= \pm1, 0)$ defines loop momentum on the internal line (j). The z_j's are the Feynman parameters to combine six propagators. The numerator function $F_\mu(D)$ is explained later.

Next we diagonalize the denominator function with respect to loop momenta ℓ_1 and ℓ_2 and perform integration. The result is,

$$\Gamma_\mu = \frac{1}{(4\pi)^n}\int\prod dz_j\delta(1-\sum_j z_j)\frac{\Gamma(6-n)}{(\mathrm{detU})^{n/2}}F_\mu(D)\frac{1}{(V-i\varepsilon)^{6-n}}, \quad U_{s,t} = \sum_{j=1}^{6} z_j\eta_s(j)\eta_t(j). \quad (2)$$

Where U is well known 2×2 matrix, composed of Feynman parameters (z_j). $V(z_j, m_j, q_j)$ is the denominator function. The arguments are easily extended to the case with five-propagators (diagrams with four-point coupling).

In order to generate the numerator function we use the following differential integral operator D_μ.

$$\frac{p_j^\mu}{(p_j^2 - m_j^2)} = D_j^\mu \frac{1}{(p_j^2 - m_j^2)}, \quad \text{where} \quad D_j^\mu \equiv \frac{1}{2}\int_{m_j^2}^{\infty} dm_j^2 \frac{\partial}{\partial q_{j\mu}} \tag{3}$$

The operator D_j^μ generates momentum p_j^μ on the internal line (j). By operating D_j^μ to the denominator function V, we get the following expression.

$$D_j^\mu \frac{1}{V^m} = \frac{Q'_{j\mu}}{V^m}, \quad Q'_j = q_j - \frac{1}{\text{detU}}\sum_{i=1} z_i B_{ij} q_j, \quad B_{ij} = \sum_{s,t} \eta_s(i)\eta_t(j) U_{st}^{-1}\text{detU} = B_{ji} \tag{4}$$

Similarly we can generate the higher order term.

$$D_i^\mu D_j^\nu \frac{1}{V^m} = \frac{Q'_{i\mu}Q'_{j\nu}}{V^m} + \left(-\frac{1}{2\text{detU}}\right)\frac{g_{\mu\nu}}{(m-1)}\frac{B_{ij}}{V^{m-1}} \tag{5}$$

We can also derive higher order terms such as $D_i^\mu D_j^\nu D_k^\lambda(1/V^m)$ and so on, however, Eq.(5) is sufficient in our case. We can verify the equivalence of the above method and the well known method of shifting momentum in the numerators. The correspondence between two methods are symbolized as follows.

$$\ell^0 \rightarrow \frac{\{1, Q'_{i\mu}, Q'_{i,\mu}Q'_{j,\nu}, \cdots\}}{V^m}, \quad \ell_i\ell_j \rightarrow \left(-\frac{1}{2\text{detU}}\right)\frac{g^{\mu\nu}}{(m-1)}\frac{B_{ij}}{V^{m-1}} \tag{6}$$

Next step is to extract the (g-2) factor by using projection operator, from the photon muon vertex Γ_μ. The quantity (g-2) is given as follows. (m_0=muon mass)

$$\begin{aligned}(g{-}2) &= \lim_{q^2\to 0}\frac{m_0}{p^4 q^2}\text{Tr}\left(\Gamma_\mu \text{Proj}(\mu)\right) \\ \text{Proj}(\mu) &= \frac{1}{4}(\not p - \frac{1}{2}\not q + m_0)\{m_0\gamma_\mu(p.p) - (m_0{}^2 + \frac{q.q}{2})p_\mu\}(\not p + \frac{1}{2}\not q + m_0),\end{aligned} \tag{7}$$

where we set momentum of incoming μ^-, outgoing μ^+ and incoming photon, as $(p{-}q/2)$, $(p{+}q/2)$ and q, respectively.

Next step is the regularization of UV-divergence. By adopting n-dimensional regularization method, any integrand F of Feynman parameter integration is regarded as a function of $\varepsilon = 4 - n/2$. Starting from the function $F(\varepsilon)$, we adopt two methods for regularization. The first method is well known one and extract $1/\varepsilon$ singularity from $F(\varepsilon)$ when one of Feynman parameters approaches 0, $(x \to 0)$. Another one is called LE-method explained below [18]. We compare two methods and conclude LE-method is simpler and more efficient than the first one if the accuracy of numerical integration is increased considerably.

First we explain the first method. In order to extract UV-divergence, we use the following procedure. First we transform the Feynman parameters $(z_1, z_2, \cdots, z_6)$ into the appropriate [0,1] variables (x, y, u, v, w) depending on the topology. The key point is to factorize the function detU$=x \times u(x, \cdots)$, where $u(0, \cdots) \neq 0$. Then it is shown that the UV-part in dimensional regularization comes from the following formula. We show the example of vertex type renormalization. We set $n = 4 - 2\varepsilon$, $\varepsilon > 0$ for UV-regularization.

$$I = \int_0^1 dx x^{\varepsilon-1}F(x,\varepsilon) = \frac{1}{\varepsilon}F(0,0) + \frac{\partial F(0,0)}{\partial\varepsilon} + \int_0^1 \frac{F(x,0) - F(0,0)}{x}dx \tag{8}$$

In the case where there is self-energy type diagram, we need the formula $\int_0^1 dx x^{\varepsilon-2} F(x,\varepsilon)$ in addition to Eq.(8). So it becomes more complex.

The second method named LE is applicable for both UV- and IR-regularization and obtaining finite values [18]. Space time dimension n is set to be $(4-2\varepsilon)$ for UV-case or $(4+2\varepsilon_R)$ for IR-case. In both cases, $\varepsilon > 0, \varepsilon_R > 0$.

After dimensional regularization method was introduced [21], the analyticity with respect to ε was discussed extensively. It is shown that the Feynman amplitude is a meromorphic function of ε [21] [22]. This is a key point to utilize the LE-method to the Feynman amplitude. Numerically, we calculate $G(i) = \int F(z_j,\varepsilon) \prod dz_j$ when $\varepsilon = \varepsilon(i)$. $(i = 1, 2, \cdots M)$. We set $\varepsilon(i) = 1/\alpha^{i+14}$ by taking relevant value α. According to the analyticity, we can expand G and truncate it at $O(\varepsilon^{M-2})$.

$$G(i) = c_{-1}\frac{1}{\varepsilon(i)} + c_0 + c_1\varepsilon(i) + \cdots + c_{M-2}\varepsilon(i)^{M-2} = \sum_{j=-1}^{M-2} c_j\{\varepsilon(i)\}^j \qquad (i = 1, 2, \cdots M) \tag{9}$$

From this formula, we get $\{c_j\}$ by multiplying the inverse of $M \times M$ matrix A, whose element is $A(i,j) = \{\varepsilon(i)\}^j$, $(i = 1, \cdots M,\ j = -1, 0, \cdots M-2)$, to M-component vector $G(i)$. Practically we set $M \sim 15$ and $\alpha \sim 1.15$.

As for counter terms, GRACE has a library of renormalization constants at one-loop level. We make use of this library for 70-diagrams composed of counter terms. We use FORM [20] to generate Fortran source, according to the above formulation.

4. Numerical Calculation

The final step to get the value (g-2) is the numerical integration over Feynman parameters. We employed trapezoidal rule with Double Exponential (DE) transformation method [23]. It is very powerful if the integrand has singular behavior at the edge of the integration domain. Speed of convergence is accelerated by the DE transformation,

$$I = \int_0^1 dx f(x) \ \rightarrow\ x = \phi(t) = \frac{1}{2}\left\{1 + \tanh\left(\frac{\pi}{2}\sinh(t)\right)\right\} \tag{10}$$

The maximum dimension of multiple integration is five. We apply DE-method to any integration parameter involved. As we need the accuracy of 7~10 digits to see the cancellation of UV, the adaptive Monte Carlo method is not suitable for two-loop calculation.

In order to ensure the validity of our results, we impose several conditions given below.

(i) Well known QED two-loop value is reproduced.
(ii) UV-divergence is cancelled.
(iii) IR-divergence is cancelled.
(iv) The result is independent of Non-linear gauge parameters.
(v) In some cases (examples: topology 4,5-a,5-b,7,9-1,9b and 10) , we perform loop-integrations (ℓ_1, ℓ_2) in turn, (we call it successive method) and obtain the same results.

In all these cases, if we have plural methods to evaluate, we compare the numerical values to ascertain the validity. If all these five conditions are cleared, we can insist on the validity of our calculation.

We demonstrate how the conditions are cleared by showing the examples.

- **Reproduction of QED** two-loop value. (Unit $=(\alpha/\pi)^2$)
  Analytic expression -0.328478966 : our value -0.328478911

- **UV-cancellation** in Feynman gauge (FG) among the diagrams.
  Examples of a group of diagrams are shown in Fig.(2)

In this case, total 13-diagrams and 1-counter term make a group to cancel UV-divergence. As you can read from the table, the cancellation is marvelous, up to almost 10 digits.

Table 2. Sample of UV-cancellation in FG

Topology	Diagram No.	Value (unit 10^{-11})
3	149	1.679393199868
3	153	0.490572808433
3	157	0.055640154059
3	161	0.055640154059
3	171	-0.021530682050
3	172	-0.006289394980
3	173	-0.021530682050
3	174	-0.006289394980
8-a	1517	-0.166093750630
8-a	1518	-0.048518165797
8-b	1578	-0.166093750630
8-b	1579	-0.048518165797
7	1671	-2.575344486333
CT-vtx	1781	0.778962156811
	Sum	-0.000000000017

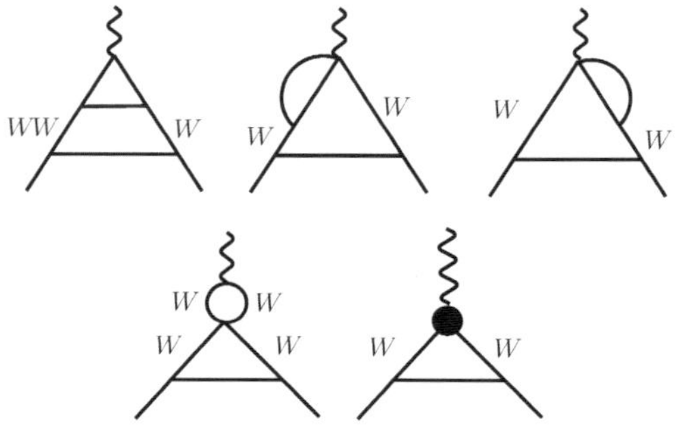

Figure 2. Sample diagrams with W

- **IR-cancellation** is also checked by using LE-method. Among two-loop diagrams, the 8 diagrams in Fig.(3) have IR-divergence. The diagrams with CT also have IR-divergence through $\delta Z_W^{1/2}$ (19-diagrams) and $\delta Z_\mu^{1/2}$ (28-diagrams).

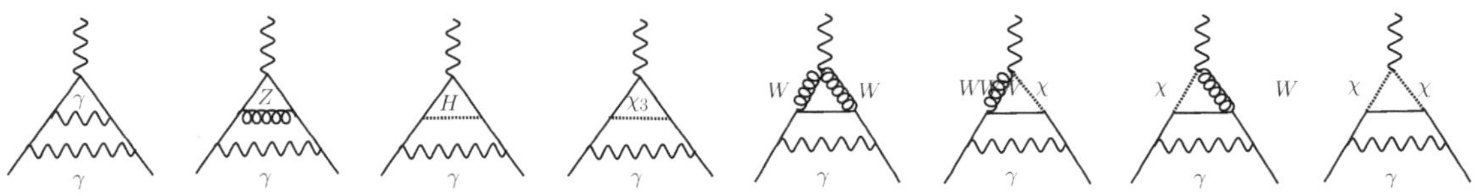

Figure 3. IR-cancellation among diagrams with W and $\chi^{\pm}$ bosons

In LE-method, the IR-divergence is proportional to $C_{IR}^{(2)} = (-1/\varepsilon_R - 2\gamma + 2\ln(4\pi))$, $\varepsilon_R = (n/2 - 2)$. It is easily shown that the IR-divergence coming from $\delta Z_W^{1/2}$ cancels among the 19-CT-diagrams. As for the diagrams with $\delta Z_\mu^{1/2}$, IR-divergence is cancelled by the corresponding two-loop diagrams. We show coefficients of $C_{IR}^{(2)}$ in Table 3. The correspondence between small photon mass (λ) method and LE-method for IR-regularization is checked in the case of QED ladder diagram. Analytic value of a coefficient of $\ln(\lambda^2/m_\mu^2)$ in unit of $(\alpha/\pi)^2$ is (1/4) [5]. It is 0.249999998 by our calculation using small photon mass. In LE-method, the coefficient of $C_{IR}^{(2)}$ becomes 0.249999999. We understand the correspondence between $\ln(\lambda^2/m_\mu^2)$ and $C_{IR}^{(2)}$ is established.

Table 3. Cancellation of IR-divergence (unit 10^{-11})

Diagram Type	$\delta Z_\mu^{1/2}$ Vertex CT	$\delta Z_\mu^{1/2}$ Self CT	two-loop diagram	sum
Neutral Type			diagram	
γ	-134887.2756×3	134887.2756×2	(a) +134887.2755	-1.6×10^{-4}
$Z - boson$	0.2152351626×3	-0.2152351626×2	(b) -0.2152351628	-1.7×10^{-10}
$Higgs, \chi_3$	$1.66293755 \times 10^{-6} \times 3$	$-1.66293755 \times 10^{-6} \times 2$	$-1.66293757 \times 10^{-6}$	-2.5×10^{-14}
Charged type			diagram(c)~(h)	
$W^\pm, \chi^\pm$	-0.4312728113299	− − −	0.4312728113330	3.1×10^{-12}

As we show in Table 3, no IR-divergence remains in the final expression.

- **Non-linear Gauge (NLG)** parameter independence is also checked. Non linear gauge was introduced to reduce the number of diagrams, particularly containing boson-boson couplings [24]. Here, we adopt NLG to check the validity of our calculation. The gauge fixing Lagrangian is constructed as,

$$\mathscr{L}_{GF} = -\frac{1}{\xi_W} F^+ F^- - \frac{1}{2\xi_Z}(F^Z)^2 - \frac{1}{\xi}(F^A)^2 \tag{11}$$

where

$$\begin{aligned} F^\pm &= \left(\partial^\mu \mp ie\tilde{\alpha}A^\mu \mp i\frac{ec_W}{s_W}\tilde{\beta}Z^\mu\right) W_\mu^\pm + \xi_W \left(M_W \chi^\pm + \frac{e}{2s_W}\tilde{\delta}H\chi^\pm \pm i\frac{e}{2s_W}\tilde{\kappa}\chi_3\chi^\pm\right) \\ F^Z &= \partial^\mu Z_\mu + \xi_z \left(M_Z \chi_3 + \frac{e}{2s_W c_W}\tilde{\varepsilon} H \chi_3\right) \\ F^A &= \partial^\mu A_\mu \end{aligned} \tag{12}$$

Here, $\tilde{\alpha}, \tilde{\beta}, \tilde{\delta}, \tilde{\varepsilon}$ and $\tilde{\kappa}$ are non-linear gauge parameters specific to this formalism. The parameters s_W and c_W are the *sine* and *cosine* of Weinberg angle θ_W.

In our calculation we set $\xi = \xi_W = \xi_Z = 1$ to make the gauge boson propagators simple. As an example of cancellation of NLG parameters, we show the UV-divergent part proportional to $\tilde{\alpha}^2$. The 53 diagrams make a group and we show sum of a coefficient of $C_{UV}^{(2)} = (1/\varepsilon - 2\gamma + 2\ln(4\pi))$ for the group of the same topological type in the Table 4. By summing up these contributions, 12 digits cancellation is confirmed even in 4 to 5 dimensional integration over Feynman parameters.

(*) Each contribution is calculated in quadruple precision method and has more effective digit than shown in the table.

Table 4. UV-part ($\tilde{\alpha}^2$-term)

Type	Number of diagrams	Coeff. of $C_{UV}^{(2)}$ in unit $(\alpha/\pi)^2$
2a	4	5.994906222E-08
2b	4	5.994906222E-08
3	7	-9.325381915E-06
5a	13	1.398809747E-07
5b	13	1.398809747E-07
8a	2	4.422894730E-06
8b	2	4.422894730E-06
9a	4	3.996606459E-08
9b	4	3.996606459E-08
sum(*)	53	-6.048741686E-18

- **Successive method** is also applied to two-loop diagrams with self energy type two-point function. An example is diagram with $(\gamma - \gamma)$ or $(\gamma - Z)$ vacuum polarization type diagram.

We decompose the renormalization constants $\delta Z^{1/2}_{AA}, \delta Z^{1/2}_{ZA}, \delta Z^{1/2}_{AZ}$ etc. into components according to the particles involved in the loop. By adding the corresponding counter term to one loop unrenormalized two-point function, we integrate over the first loop momentum ℓ_1. The result is finite renormalized two-point function $\Pi_R(\ell_2)$. The function Π_R contains the second loop momentum ℓ_2, however, the divergent part of ℓ_2 integration does not contribute to (g-2). We use this alternative method to reconfirm the results obtained by the methods given in section 3.

5. Finite contribution from restricted type of diagrams

The publicly accepted value of ELWK two-loop correction is classified according to types of diagrams. For example, fermion loop contribution with or without Higgs boson and purely bosonic contribution are discussed separately. As an example, we summarize the naive fermion loop contribution without any approximation. The diagrams we consider are shown in Fig.4. As is shown in Fig.4, we consider diagrams with both fermion loop and neutral boson.

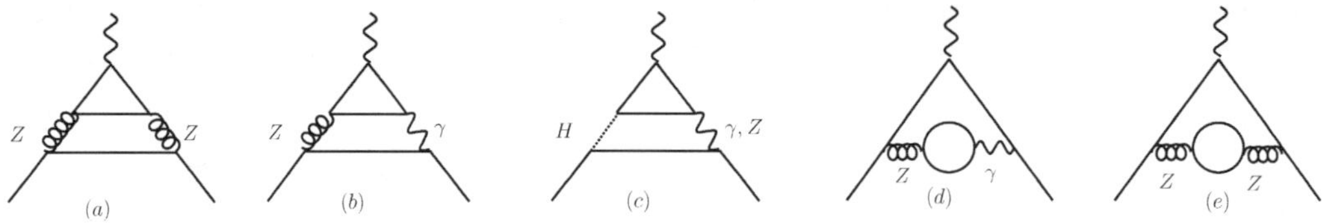

Figure 4. Fermion loop diagrams

The contribution from fermion triangle diagrams Fig.(4-a,b) comes from VVA-anomaly. If lepton and quark mass within the same generation are degenerate, VVA-anomaly cancellation works so that the contribution becomes zero. So fermion mass difference is crucial to get the value. The $\ln(M_z^2)$-dependence cancels within one generation, because of no-anomaly condition. Czarnecki et al.[6] give the approximate formula for diagram Fig.(4-b). Taking the same parameters (see below), we compare their values and our exact results in Table 5.

Table 5. Comparison of Fig.(4-b) (unit 10^{-11})

lepton and quark	Ref.[6] approximate value	Our exact value
electron, up,down	-3.9984	-4.0272
muon, charm,strange	-4.6524	-4.6426
τ, top, bottom	-8.1854	-8.1940
Total	-16.8362	-16.8638

Although each fermion contribution is quite different, the coincidence becomes pretty good by taking the sum in each generation. The difference comes from the common constant terms which cancels by summing up within each generation, due to the anomaly free condition. We also show the preliminary value of other diagrams Fig.(4-a,c,d,e) in Table 6.

Table 6. Finite value Fig(4-a,c,d,e)

Type	Value $\{10^{-11}\}$
4-a	-0.0494
4-c	-1.8718
4-d	-0.1803
4-e	1.5842

We use the following fermion and boson masses in the calculation in unit GeV. $m_\mu = 105.658389\times10^{-3}, m_e = 0.51099906\times10^{-3}, m_\tau = 1.7771, m_u = 0.3, m_c = 1.5, m_t = 173.21, m_d = 0.3, m_s = 0.5, m_b = 4.5, M_W = 80.22, M_Z = 91.187, M_H = 125.09$

6. Discussion and Comment

We develop the system to calculate the full ELWK two-loop corrections to $(g\text{-}2)$. Unfortunately it is not quite completely automatic system yet. We can produce Fortran source automatically. The work we need beforehand is only to prepare several files which only depend on the type of the topology of diagrams as we explained in section 2.

If we adopt the well known first method to extract UV-divergent part, it is crucial to introduce the most suitable transformations from (z_j) to [0,1] integration variables (x, y, u, v, w). In the case of Linear Expansion method (LE), however, the choice of integration variables is not sensitive to get the reliable results.

Further comments are given on the regularization method. Firstly, we introduce n-dimensional regularization method and take out UV-divergence $(1/\varepsilon)$, when one of the integration variables is approaching 0 $(x \to 0)$. In order to perform this procedure, we need rather complex operations including differentiation of the amplitudes, etc. As a result, it makes Fortran source lengthy so that CPU-time increases extensively. The second prescription to regularize both UV- and IR-divergences is LE-method, briefly explained in section 3. This method decreases the number of operation drastically. It is sufficient to define the quantity as function of $\varepsilon(= 2 - n/2)$. We only need to treat Dirac matrices and various vectors appeared in the numerator, in n-dimension. This is easily done by using symbolic manipulation system such as FORM. The operation is simple and we can make use of the resultant short sources for both UV-$(\varepsilon > 0)$ and IR-$(\varepsilon_R = -\varepsilon > 0)$ regularization and also to get finite results. By comparing the results of various methods mentioned in the previous sections, we conclude that LE-method is the most simple and reliable method, at this moment. In order to get reliable physical value by this method, accurate numerical integration over Feynman parameters is inevitable. The DE-method introduced in section 4 is the suitable candidate.

By using these technical approaches mentioned above, we clear almost all the constraints given in section 4. Namely, (i) reproduction of QED values, (ii)(iii) UV-and IR-divergences are cancelled, (iv) the result is independent of NLG-parameters. We show some samples how they are cleared. We are now recalculating and rechecking the constraints under the best computer circumstances. The final value will be given soon. We expect ongoing work provides the fruitful foundation to formulate PNQFT (Perturbative Numerical Quantum Field Theory) mentioned in section 1.2.

Acknowledgments

We would like to thank Prof.T.Kaneko for his important contribution to construct the framework of calculation at the early stage of this work. We also wish to thank Prof.K.Kato and

Prof.F.Yuasa for discussion. Last but not least, we express our deep appreciation to late Prof.Y.Shimizu for his continual encouragement. This research is partially supported by Grant-in-Aid for Scientific Research (15H03668,15H03602) of JSPS.

References

[1] Yuasa F et al. 2000 *Prog. Theore. Phys. Suppl.* **138** 18
Belanger G, Boudjema F, Fujimoto J, Ishikawa T, Kaneko T, Kato K and Shimizu Y 2006 *Phys.Rept.* **430** 117
[2] Aoyama T, Hayakawa M, Kinoshita T and Nio M 2012 *Phys.Rev.Lett.* **109** 111808
[3] Kukhto T V, Kuraev E A, Schiller A and Silagadze Z K 1992 *Nucl.Phys.* **B371** 567
[4] Kaneko T and Nakazawa N 1995 *Proc.of AIHENP(Artificial Intelligence for High Energy and Nuclear Physics*,(Pisa,Italy) 173
[5] Karplus R and Kroll N M 1950 *Phys.Rev.***77** 536
Sommerfield C M 1957 *Phys.Rev.***107** 328; 1958 *Annals of Physics* **5** 26
Petermann A 1957 *Hel.Phys.Acta***30** 407; 1958 *Nucl. Phys.* **5** 677
[6] Czarnecki A, Krause B and Maciano W J 1995 *Phys.Rev.* **D52** R2619
Czarnecki A, Krause B and Marciano W J 1996 *Phys.Rev.Lett.* **76** 3267
Czarnecki A, Marciano W J and Vainshtein A 2003 *Phys.Rev.* **D67** 073006 Erratum *ibid.* 2006 *Phys.Rev.* **D73** 119901
[7] Gribouk T and Czarnecki A 2005 *Phys.Rev.* **D72** 053016
[8] Marquard P 2015 *Nuclear and Particle Physics Proceedings* **260** 107
[9] Bennett G W *et al.* 2002 *Phys.Rev.Lett.* **89** 101804
Erratum *ibid* 2002 Phys.Rev.Lett. **89** 129903
Bennett G W *et al.* 2004 *Phys.Rev.Lett.* **92** 161802
Bennett G W *et al.* 2006 *Phys.Rev.* **D73** 072003
[10] Patrignani C *et al.* (Particle Data Group) 2016 *Chin.Phys.* **C40** 100001 pp.715
[11] Davier M, Hoecker A, Malaescu B and Zhang Z 2011 *Eur.Phys.J.* **C71** 1515
[12] Prades J, de Rafael E and Vainshtein A 2009 *Advanced series on directions in high energy physics* **vol. 20** ed Roberts B L and Marciano W pp 303-317
[13] Jackiw R and Weinberg S 1972 *Phys.Rev.* **D5** 2396
Altarelli G, Cabibbo N and Maiani L 1972 *Phys.Lett.* **B40** 415
Bars I and M Yoshimura 1972 *Phys.Rev.* **D6** 374
Fujikawa K, Lee B W and Sanda A I 1972 *Phys.Rev.* **D6** 2923
[14] Gnendiger C, Stöckinger D and Stöckinger-Kim H 2013 *Phys.Rev.* **D88** 053005
Heinemeyer S, Stöckinger D and Weiglein S 2004 *Nucl.Phys.* **B699** 103
[15] Mohr P J, Taylor B N and Newell D B (CODATA Group) 2012 *Rwv.Mod.Phys.* **84** 1527
[16] Gohn W 2016 The Muon g-2 experiment at Fermilab *Preprint* arXiv:1611.04964[hep-ex]
[17] Otani M (E34 collaboration) 2015 *JPS Conf.Proc.* **8** Status of the Muon g-2/EDM Experiment at J-PARC(E34) 025008
[18] de Doncker E, Yuasa F and Kurihara Y 2011 *Proc., 3rd CPP Workshop on Computational particle physics (CPP 2010)* (Tsukuba, Japan, September 23-25,2010) p 11
de Doncker E, Fujimoto J, Hamaguchi N, Ishikawa T, Kurihara Y, Shimizu Y and Yuasa F 2012 *Jounal of Computational Science* **3** 102
de Doncker E, Yuasa F, Kato K, Ishikawa T,Kapenga J and O.Olagbemi 2017, *Preprint* arXiv:1702.04904[hep-ph]
[19] Civitanović P and Kinoshita T 1974 *Phys.Rev.* **D10** 3978, *ibid* 3991
[20] Vermaseren J A M 2000 *preprint* arXiv:0010025[math-ph]
[21] Bollini C G and Giambiagi J J 1972 *IL Nuovo Cimento B* **12** 20
' t Hooft G and M Veltman 1972 *Nucl.Phys.* **B44** 189
[22] Nakanishi N 1975 *Quantum Field Theory* (Tokyo:Baihukan) p.272 (in Japanese)
[23] Takahashi H and Mori M 1974 *Publ. RIMS Kyoto Univ.* **9** 721
[24] Boudjema Z and Chopin E 1996 *Z.Phys.Rev.* C**73** 85

IOP Conf. Series: Journal of Physics: Conf. Series **920** (2017) 012010 doi:10.1088/1742-6596/920/1/012010

1-loop effects of MSSM particles in Higgs productions at the ILC

Yusaku Kouda[1], Tadashi Kon[1], Masato Jimbo[2], Yoshimasa Kurihara[3], Tadashi Ishikawa[3], Kiyoshi Kato[4], Masaaki Kuroda[5]

[1] Seikei University, Musashino, Tokyo 180-8633, Japan
[2] Chiba University of Commerce, Ichikawa, Chiba 272-0827, Japan
[3] KEK, Tsukuba, Ibaraki 305-0801, Japan
[4] Kogakuin University, Shinjuku, Tokyo 163-8677, Japan
[5] Meiji Gakuin University, Yokohama, Kanagawa, 244-8539, Japan

E-mail: dd146101@cc.seikei.ac.jp

Abstract. Within the framework of the MSSM (Minimal Supersymmetric Standard Model), we investigate the 1-loop effects of SUSY (supersymmetric) particles on the Z,higgs production at the ILC. Three sets of the SUSY parameters are proposed which are consistent with the observed higgs mass, the muon g-2, the dark matter (DM) abundance and the decay branching ratio of B meson. In addition, We investigate the 1-loop effects on $\nu\bar{\nu}h$ production using W-fusion approximation at the ILC. We discuss on the possibility of discovering the signals consistent with SUSY as well as of experimentally distinguishing the proposed sets of SUSY parameters.

1. Introduction

The supersymmetric (SUSY) model [1] is considered as one of the promising candidates for the theory beyond the standard model. The experimental confirmation of the SUSY particles (sparticles) is an important subject of the present and future collider experiments. Since high luminosity is expected at the ILC experiments [2], we should calculate the physical observables with the accuracy better than the experimental data. We have calculated several cross sections and decay branching ratios at 1-loop level with `GRACE/SUSY-loop` system [3, 4, 5]. In this paper we report numerical results of the cross section of $e^-e^+ \to Zh$. (The same observable calculated in the previous work [6], and we confirmed almost same numerical results at the same assumption for the MSSM input parameters. We also calculeted the W fusion approximation of $e^-e^+ \to \nu_e\bar{\nu}_e h$, where the SM 1-loop correction has already been calculated with using `GRACE` [7]).

2. Selection of the MSSM parameter sets

We have considered the experimental constraints (Table 1) in selection of the typical sets of MSSM parameters. We selected three sets [19] which consistent with next constraints. We have calculated the cross section of $e^+e^- \to Zh$ at the 1-loop level in these sets. We used MicrOMEGAs [11] in the estimation of the MSSM prediction for the dark matter (DM) thermal relic density(1), [8, 9, 10]. The program package SuSpect2 [18] are used for the muon g-2

IOP Conf. Series: Journal of Physics: Conf. Series **920** (2017) 012010 doi:10.1088/1742-6596/920/1/012010

anomalous magnetic moment (2), [12, 13], the B meson rare decay branching ratio (3), [14] and the observed higgs mass (4), [15, 16]. Moreover, we set the sparticle masses which meet the recent LHC bound (5), [17].

Table 1. The experimental constraints for MSSM parameters.

Constraint	Numerical value
(1) DM thermal relic density	$\Omega h^2 = 0.1198 \pm 0.0026$
(2) muon g-2 anomalous magnetic moment	$a_\mu^{\rm exp} - a_\mu^{\rm SM} = (2.88 \pm 0.63 \pm 0.49) \times 10^{-9}$
(3) B meson rare decay branching ratio	$\mathrm{Br}(B \to \chi_s \gamma) = (3.43 \pm 0.21 \pm 0.07) \times 10^{-4}$ $\mathrm{Br}(B_s \to \mu^+\mu^-) = (3 \pm 1) \times 10^{-9}$
(4) Higgs mass	$m_h(exp) = 125.09 \pm 0.24$GeV
(5)LHC direct search of sparticles	$m_{\tilde{q}}, m_{\tilde{g}} \gtrsim 1.5$ TeV

In Table 2, we show the set which is adopted for the calculation of $e^+e^- \to \nu_e\bar{\nu}_e h$ cross section with the W fusion approximation . It is noted that the set is consistent with the constraints (1),(3),(4),(5). In other words, we have not considered the muon g-2 constraint (2) in table 2. We have shown in the previous work [19] that the anomaly (2) can only be explained by MSSM with $m_{\tilde{\chi}_1^0} \lesssim 450$ GeV. Here, we set $m_{\tilde{\chi}_1^0}$ (neutralino1 mass) = 552.8 GeV and calculated $\sqrt{s} = 500$ GeV, so the channel $e^-e^+ \to \tilde{\chi}_1^0\tilde{\chi}_1^0 h$ is not considered.

Table 2. The MSSM mass spectra of the set which we selected for the $\nu_e\bar{\nu}_e h$ cross section calculation(The units are GeV except the mixing angles θ and tanβ).

h	H	A	H^+	$\tilde{\chi}_1^0$	$\tilde{\chi}_2^0$	$\tilde{\chi}_3^0$	$\tilde{\chi}_4^0$
125.2	2000	2000	2002	552.8	588.2	614.8	1376
χ_1^+	χ_2^+	$\tilde{\ell}_1$	$\tilde{\ell}_2$	$\tilde{\nu}_\ell$	$\tilde{\nu}_\tau$	$\tilde{\tau}_1$	$\tilde{\tau}_2$
583.3	1376	601.5	651.7	641.9	646.9	589.5	662.5
$\tilde{u}_1$	$\tilde{u}_2$	$\tilde{d}_1$	$\tilde{d}_2$	$\tilde{t}_1$	$\tilde{t}_2$	$\tilde{b}_1$	$\tilde{b}_2$
5000	5000	4800	5000	1798	2508	2200	2501
θ_τ	θ_b	θ_t	M_1	M_2	M_3	μ	tanβ
1.164	1.539	1.481	585.0	1370	2500	586	30

IOP Conf. Series: Journal of Physics: Conf. Series **920** (2017) 012010 doi:10.1088/1742-6596/920/1/012010

3. Numerical Results

We show total cross sections for associate higgs particle production processes at future linear collider in Figure 1. $\sqrt{s}$ = 250 GeV is adopted for the precision verification of $e^-e+ \to Zh$. Since $\sigma(e^-e^+ \to \nu_e\bar{\nu}_e h)$ is larger than $\sigma(e^-e^+ \to Zh)$ and $\sigma(e^-e^+ \to \nu_e\bar{\nu}_e h)$ via the W-fusion are dominant contribution for $\sqrt{s} \gtrsim 500$ GeV.

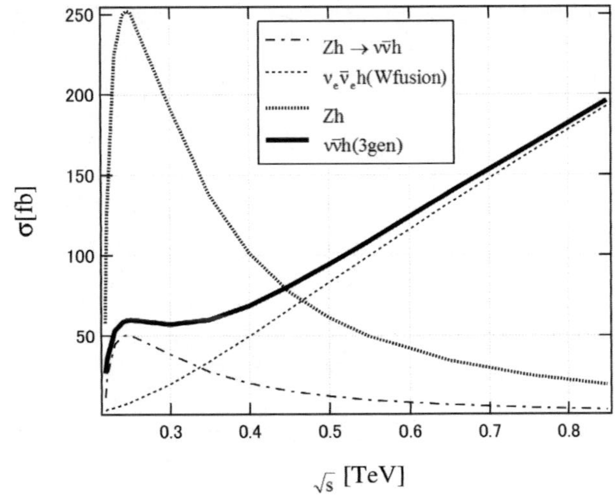

Figure 1. The $\sqrt{s}$ dependence of the total cross sections at the ILC for different final states. $\nu\bar{\nu}h$(3gen) stands for the sum of $\nu_e\bar{\nu}_e h$, $\nu_\mu\bar{\nu}_\mu h$ and $\nu_\tau\bar{\nu}_\tau h$.

For $\sqrt{s}$ = 250(500) GeV, the 1-loop correction at cosθ =0, namely, $\delta_{\rm SM} \equiv \frac{d\sigma^{\rm SM}_{\rm 1loop} - d\sigma_{\rm tree}}{d\sigma_{\rm tree}}$ = -11(9) %, and $\delta_{\rm MSSM} \equiv \frac{d\sigma^{\rm MSSM}_{\rm 1loop} - d\sigma_{\rm tree}}{d\sigma_{\rm tree}} = -9(11)\%$. We have confirmed forward-backward symmetry of the angular distribution even at the 1-loop level.

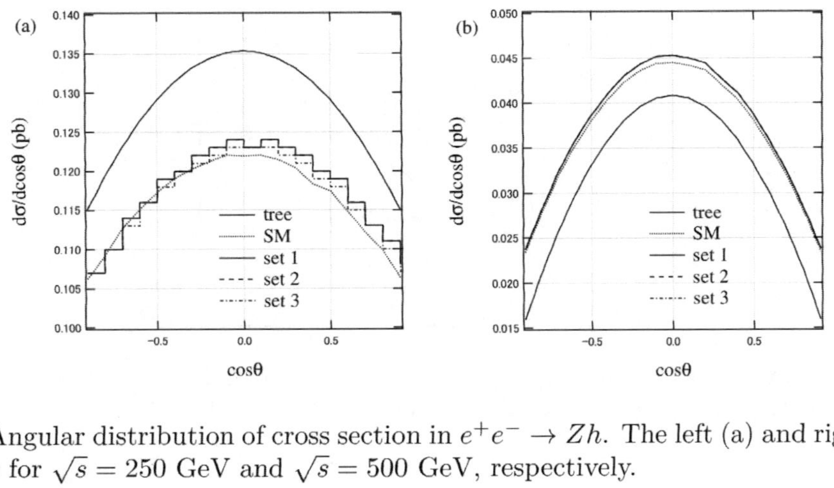

Figure 2. Angular distribution of cross section in $e^+e^- \to Zh$. The left (a) and right (b) figures show results for $\sqrt{s} = 250$ GeV and $\sqrt{s} = 500$ GeV, respectively.

We define the ratio of the differential cross sections [20],

$$\delta_{\rm susy} \equiv \delta_{\rm MSSM} - \delta_{\rm SM} = \frac{d\sigma^{\rm MSSM}_{\rm 1loop} - d\sigma^{\rm SM}_{\rm 1loop}}{d\sigma_{\rm tree}}. \quad (1)$$

We show $\delta_{\rm susy}$ in Figure 3. $\delta_{\rm susy}$ is 1~2% in the entire region and is larger than the statistical error. It means that the 1-loop contribution of the MSSM could be measured at both $\sqrt{s}$ =250 GeV and 500 GeV. In addition, the difference between set2 and set3 is also larger than the statistical error at $\sqrt{s}$=250 GeV.

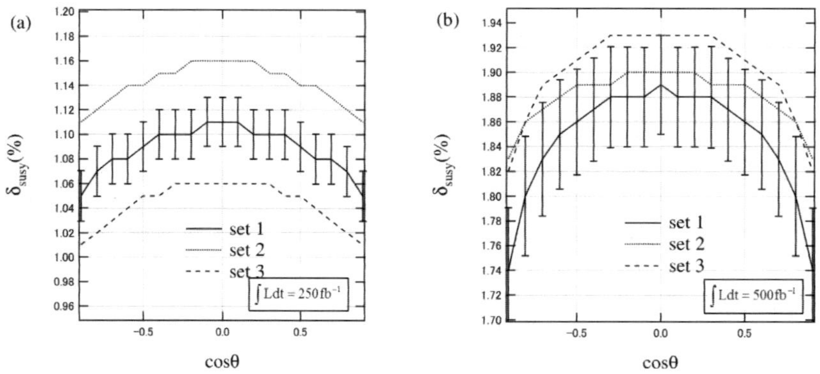

Figure 3. Correction ratio $\delta_{\rm susy}$ in $e^+e^- \to Zh$. The left (a) and right (b) figures show results for ($\sqrt{s}, L$) =(250 GeV, 250 fb^{-1}) and ($\sqrt{s}, L$)= (500 GeV, 500 fb^{-1}), respectively.

For $\sqrt{s}$ = 500 GeV, the single Higgs production $e^-e^+ \to \nu_e\bar{\nu}_e h$ with the W-fusion approximation is also calcurated. We choose the W-fusion type 239 diagrams from the full 1-loop 13793 diagrams. An example is shown in Figure4.

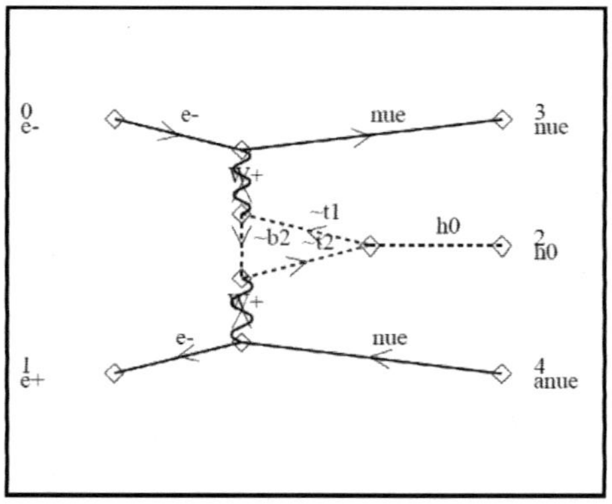

Figure 4. An example of Feynman diagrams in which the virtual W pair exists in the internal lines. stops ($\tilde{t}$) and sbottoms ($\tilde{b}$) contribute in the 1-loop.

We confirmed cancellation of ultraviolet and infrared divergences, and the photon cut-off energy independence. The Energy distribution and correction ratio $\delta_{\rm susy}$ in W fusion are shown in Figure 5. Where, E_h is single generation energy of higgs particle. The 1-loop correction at E_h

$= 150$ GeV $\delta_{\rm SM} \equiv \frac{d\sigma^{\rm SM}_{\rm 1loop} - d\sigma_{\rm tree}}{d\sigma_{\rm tree}} = -36\%$ and $\delta_{\rm MSSM} \equiv \frac{d\sigma^{\rm MSSM}_{\rm 1loop} - d\sigma_{\rm tree}}{d\sigma_{\rm tree}} = 22\%$. In the entire region, $\delta_{susy} = 11\% \sim 17\%$. It is larger than the error of the Monte Carlo integration. It means that the MSSM signal is verifiable, but this results are preliminary and necessary to be scrutinized more.

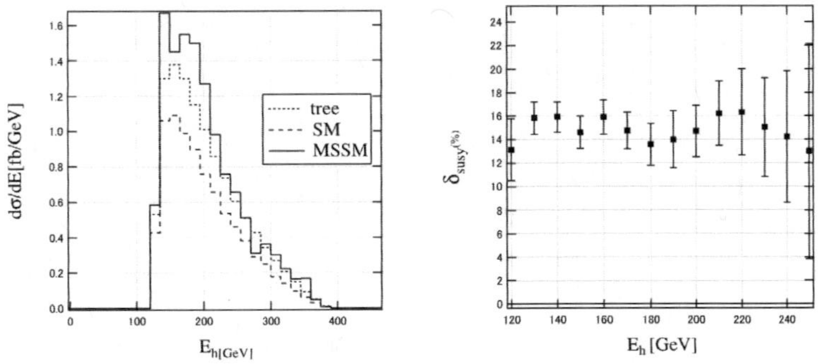

Figure 5. The energy distribution and correction ratio $\delta_{\rm susy}$ in $e^-e^+ \to \nu_e\bar{\nu}_e h$. The left figure shows energy distribution at $\sqrt{s} = 500$ GeV and The right figure shows the correction ratio $\delta_{\rm susy}$.

4. Summary

We selected sets that are consistent with higgs mass, B physics, DM relic density, LHC direct search of sparticles, and muon g-2(only in Zh production). We have probed statistical significance at the ILC. In the results of Zh, the MSSM 1-loop effects are statistically significant at both of $\sqrt{s}$= 250 GeV and 500 GeV. At $\sqrt{s}$ =250 GeV, three sets are distinguishable. The results of W fusion approximation in e$^-$e+ $\to \nu_e\bar{\nu}_e h$ are preliminary, but the 1-loop effect of supersymmetry would be verifiable at $\sqrt{s} = 500$ GeV in the future linear collider.

5. References

[1] S. P. Martin, Adv. Ser. Direct. High Energy Phys. **18**, 1 (1998), hep-ph/9709356v7 (2016).
[2] H. Baer, T. Barklow, K. Fujii, Y. Gao, A. Hoang, S. Kanemura, J. List, H. E. Logan, A. Nomerotski, M. Perelstein, et al. (2013), arXiv:1306.6352.
[3] J. Fujimoto, T. Ishikawa, M. Jimbo, T. Kaneko, K. Kato, S. Kawabata, T. Kon, M. Kuroda, Y. Kurihara, Y. Shimizu, and H. Tanaka, Comput. Phys. Commun., **153**, 106 (2003).
[4] J. Fujimoto, T. Ishikawa, Y. Kurihara, M. Jimbo, T. Kon, and M. Kuroda, Phys. Rev. D, **75**, 113002 (2007).
[5] M. Jimbo, T. Kon, Y.Kouda,. M Ichikawa, Y Kurihara, T Ishikawa, K Kato, and M. Kuroda, (2017) arxiv:1703.07671 [hep-ph]
[6] J. Cao, C. Han, J. Ren, L. Wu, J. Yang, M. Jin and Y. Zhang,, Chin. Phys. **C40** 113104 (2016)
[7] G. Blanger , F. Boudjema , J. Fujimoto , T. Ishikawa , T. Kaneko, K. Kato, and Y. Shimizu, Physics Letters B 559 (2003)
[8] A. Ibarra, A. Pierce, N. R. Shah, and S. Vogl, Phys. Rev. D, **91**, 095018 (2015).
[9] J. Ellis, K. A. Olive, and J. Zheng, Eur. Phys. J., **C74**, 2947 (2014).
[10] K. A. Olive, PoS, **PLANCK2015**, 093 (2015).
[11] G. Bélanger, F. Boudjema, A. Pukhov, and A. Semenov, Comput. Phys. Commun., **149**, 103 (2002).
[12] A. Hoecker and W. J. Marciano, The Muon Anomalous Magnetic Moment, in Particle Data Group, Chin. Phys., **C38**, 090001 (2014); p.649 (updated August 2013), references therein.
[13] G. C. Cho, K. Hagiwara, Y. Matsumoto, and D. Nomura, JHEP, **11**, 068 (2011).
[14] Amhis, Y. and others, FERMILAB-PUB-15-004-PPD, (2014)
[15] ATLAS Collaboration, Phys. Lett., **B716**, 1 (2012).
[16] CMS Collaboration, Phys. Lett., **B716**, 30 (2012).
[17] ATLAS Collaboration, JHEP, **09**, 176 (2014).

IOP Conf. Series: Journal of Physics: Conf. Series **920** (2017) 012010 doi:10.1088/1742-6596/920/1/012010

[18] A. Djouadi, J.-L. Kneur, and G. Moultaka, Comput. Phys. Commun., **176**, 426 (2007).
[19] Y. Kouda, T. Kon, Y. Kurihara, T. Ishikawa, M. Jimbo, K. Kato,. and M. Kuroda, Prog Theor Exp Phys doi:10.1093/ptep/ptx048 **2017 5**, 053B0 (2017).
[20] W. Hollik and C. Schappacher, Nucl. Phys., **B545**, 98–140 (1999).

IOP Conf. Series: Journal of Physics: Conf. Series **920** (2017) 012011 doi:10.1088/1742-6596/920/1/012011

Renormalisation of the Higgs sector of the NMSSM. Application to Higgs decays

Fawzi Boudjema

LAPTh, Université Savoie Mont Blanc, CNRS, B.P.110, Annecy-le-Vieux F-74941, France

E-mail: `boudjema@lapth.cnrs.fr`

Abstract. I will first recollect some memories of my collaboration with Shimizu-sensei, a collaboration that mainly covered the topic of renormalisation and loop corrections. I will then describe the renormalisation of the Higgs sector of the NMSSM, this work could be seen as a natural continuation of the long collaboration I have had with the Minami-Tateya group. This study has been done with Geneviève Bélanger, Guillaume Chalons and Vincent Bizouard. Several on-shell renormalisation schemes as well as a mixed on-shell $\overline{\rm DR}$ scheme are analysed. Using `SloopS`, a code that can perform automatic computation of one-loop corrections, the one-loop electroweak corrections to the partial widths of the pseudoscalar and scalar Higgses into lighter Higgses and/or neutralino/charginos are computed.

1. Introduction

My first correspondance with Shimizu-sensei was around 1987 when I was graduating from Sussex University, a few months earlier than the time when Norman Dombey and I where preparing the Workshop *Radiative Corrections: Results and Perspectives*[1] in Brighton, UK, in July 1989. This conference, just at the start of LEP1, turned out to be the first in the series that has come to be known as `RADCOR`. The correspondance with Shimizu was about a work on precision calculations, $e^+e^- \to \gamma\gamma$[2]. The first time we met was around 1993 in Annecy, shortly after I joined the CNRS with LAPTh. Our first collaboration was during the year-long *Physics at LEP2* CERN Workshop in 1994-1996 when I was convening the *Standard Model Processes* working group[3]. Shimizu with his Minami-Tateya group had contributed in a major way and showed the prowess of the `GRACE` system. I was not part of the `GRACE` collaboration but its through the CERN Yellow Book (LEP2) that my long discussions with Shimizu-sensei started. In the next year or two that followed, we embarked on a common project, the automation of one-loop electroweak corrections. LEP2 had revealed the importance of four-body final states where many groups competed to provide results for many channels of interest at LEP2. With the concomitant interest in the physics of a high energy e^+e^- linear collider, the project aimed at providing better precision for many processes at the Linear Collider. Many of these processes required the calculation of multi-leg processes. Prime among these processes were Higgs production mechanisms $e^+e^- \to \nu\bar{\nu}H, t\bar{t}H, e^+e^-H, ZHH, \nu\bar{\nu}HH$ and .. 4-fermion process that are needed to go beyond the important resonant contributions from $e^+e^- \to W^+W^-$. Before embarking on this (then) titanic programme, there was no full one-loop calculations to $2 \to 3$ and $2 \to 4$ processes. We wanted to push the theoretical challenge and set these processes as targets. From the end of 1997 the exchanges were intense, as were the visits with which my love for Japan grew tremendously. The first concrete aspect aimed at finding a powerful internal check of the code. Of course we had to make sure that the renormalisation procedure was implemented correctly by checking for each process that the corrections were ultra-violet finite. But much more powerful was the check on gauge invariance and/or gauge-parameter independence on all physical amplitudes. The problem is that we wanted efficiency, however reverting to the usual ξ 't Hooft-Feynman gauges would have been intractable. The "longitudinal" terms in the vector boson propagators (which are absent in the much used Feynman gauge) not only introduce a high (superficial) degree of divergence in the intermediate steps of the computation but, perhaps worse, generate loop integrals with very high tensorial structures. Reduction of these (very) high rank tensors on the basis of scalar integrals is a tremendous and expensive task. The idea therefore was to have the gauge parameter dependence (in the intermediate stages of the calculations) in each amplitude in a simple polynomial form while keeping the analytic structure of the loop integral evident with only the physical masses showing up in the corresponding diagrams. We used a non-linear gauge fixing[4, 5] with no less than 5 gauge parameters: $\tilde{\alpha}, \tilde{\beta}, \tilde{\delta}, \tilde{\kappa}, \tilde{\epsilon}$. Denoting $A, Z, W^{\pm}, H$ for, respectively, the photon, the Z boson, the W and the Higgs boson, this non-linear gauge-fixing writes

$$\begin{aligned}\mathcal{L}_{GF} &= -\frac{1}{\xi_W}|(\partial_\mu - ie\tilde{\alpha}A_\mu - igc_W\tilde{\beta}Z_\mu)W^{\mu+} + \xi_W\frac{g}{2}(v+\tilde{\delta}h + H + i\tilde{\kappa}\chi_3)\chi^+|^2 \\ &\quad -\frac{1}{2\xi_Z}(\partial.Z + \xi_Z\frac{g}{2c_W}(v+\tilde{\epsilon}H)\chi_3)^2 - \frac{1}{2\xi_\gamma}(\partial.A)^2 \end{aligned} \qquad (1)$$

I retrieved an e-mail from Shimizu-sensei dated 06/10/1997, exactly 19 years ago to the day!, stating that this step of the set-up for the full one-loop computation had been completed and I was invited to KEK so that the implementation of the non-linear gauge be completed in the `GRACE` system in view of the evaluation of the full one-loop amplitudes in the electroweak theory. This non-linear gauge was implemented in `GRACE-Loop` together with the full set of counterterms. It took a couple of years for the system to be ready for the first challenging process $e^+e^- \to \nu\bar{\nu}H$. The first preliminary results[6] were shown at `RADCOR 2002/LOOP and LEGS 2002` in September by Fujimoto-san in Kloster Banz (Germany). We were all a bit nervous, dreading to be scooped. From Germany Fujimoto was reporting by phone directly to Shimizu who in turn reported to me (from Japan!) by e-mail. In a message I recovered from my e-mail archive, the evening following Fujimoto presentation, Shimizu wrote "*.... He (Fujimoto-san) said in last evening session Tarasov has given a talk on nu-nu-Higgs 1-loop calculations but without showing any numbers. ..... It seems that now the world competition is becoming very tough like in 1995 on W-pair for LEP-2. Sayonara,Shimizu Y* ". The mail came with a caring post-scriptum "*Though you will certainly get the true explanation tomorrow, I tried to tell my guess so that you can have a good sleep tonight!*
Our Physics Letters B[7] on $\nu\bar{\nu}H$ appeared soon after. We were not scooped. At that point the collaboration was in full swing, with frequent visits to Japan and for our Minami-Tateya friends to France. The items on our list of priority processes were ticked one after the other[8, 9, 10, 11, 12, 13, 14]. This period corresponded with one of the happiest period in my life, my eldest daughter being born in 2001. The picture of Shimizu-sensei playing with Noor brings some sweet memories to me. At the same time as `GRACE-1Loop` was developing, Minami-Tateya was also developing the SUSY version and we exchanged also regarding this progress. Later on, it was decided that one should perform a full renormalisation of the MSSM and that the Annecy group should perform independent checks. In Annecy with the help of Andrei Semenov we decided to exploit `LANHEP`[16] as a model file builder to the `FeynArts, FormCalc,LoopTools`[17] packages in order to have an independent system which we called `SloopS`[18, 19, 20, 21]. In retrospect, if I go through my interaction with Shimizu-sensei, the first encounter was on a QED reaction controlled essentially with one parameter, α, then we

IOP Conf. Series: Journal of Physics: Conf. Series **920** (2017) 012011 doi:10.1088/1742-6596/920/1/012011

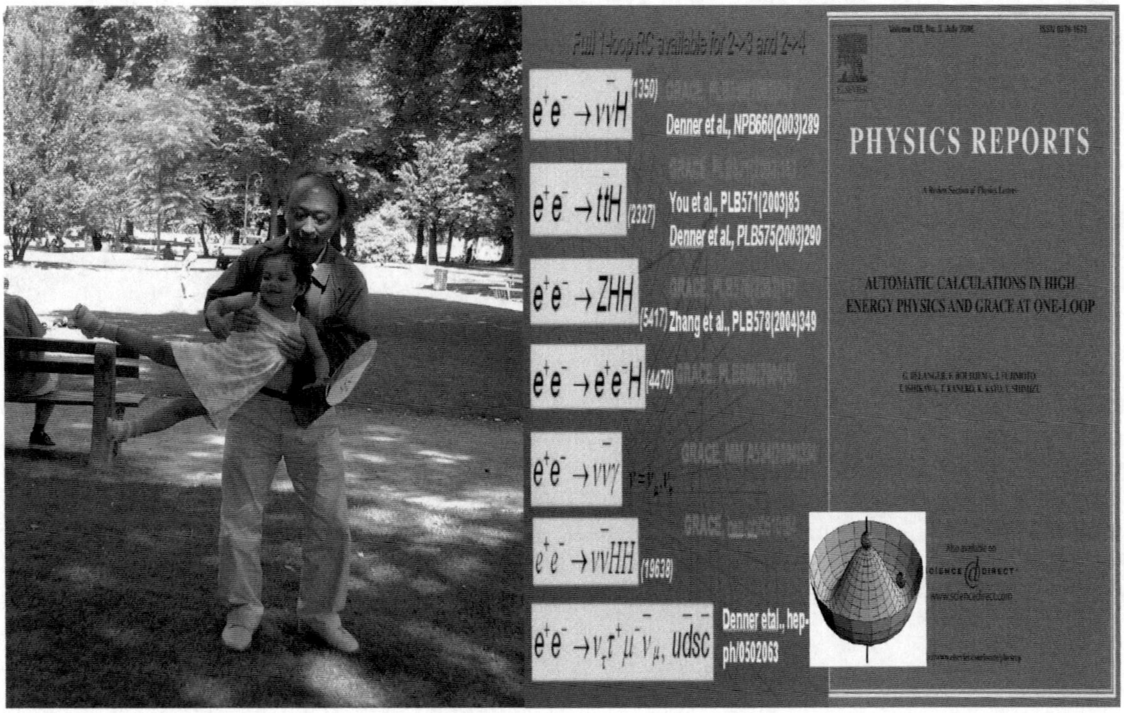

Figure 1. In *full swing*, Shimizu playing with Noor at the time we were writing our Physics Reports[15]

moved to a theory with a few parameters ($SU(2) \times U(1)$), specialising to multi-leg processes and then to a multi-parameter theory like the MSSM. Working with Shimizu-sensei, precision physics was an experience of beauty in different levels of complexity! Being in Japan, I can not help making the connection between our multi-leg calculations in multi-parameter theories with the multiplication of legs and arms within the multitude of statues in Sanjusangendo! Dealing with multi-parameter theories, eventually with multi-arm processes is not the same as what Gudrun referred to (for QCD N(N)LO) processes), mass production of cookies! It's rather recreating the subtlety in the difference contained in the 1000 figures demultiplied through arms (and legs) of the Sanjusangendo hall.

I will now report on one such theory that has even more parameters than the MSSM and the issues one has to face in renormalising this theory. The renormalisation of all the sectors of the NMSSM has been completed in `SloopS`[22, 23] here I will report on the Higgs sector. I will concentrate in particular on the definition/reconstruction of the parameters and the passage from the parameters at the Lagrangian level to the more physical parameters that may be chosen as inputs instead, namely the masses of the particles of the theory. The issue of mixing between the fields is critical, as is the issue of scheme and scale dependence.

2. The NMSSM

The NMSSM[24] provides a natural explanation for the scale of the higgsino parameter, μ by relating it to the vev of a scalar singlet, thus solving the little hierarchy problem of the MSSM. It makes it also much easier to produce a Higgs with a mass of 125GeV as observed at the LHC. For the MSSM such a mass requires large values in the stop sector which in turn creates an unbearable unease with naturalness. The NMSSM contains three Higgs superfields : two

$SU(2)_L$ doublets $\hat{H}_u$ and $\hat{H}_d$, as in the MSSM, and one additional gauge singlet $\hat{S}$

$$\hat{H}_u = \begin{pmatrix} \hat{H}_u^+ \\ \hat{H}_u^0 \end{pmatrix}, \quad \hat{H}_d = \begin{pmatrix} \hat{H}_d^0 \\ \hat{H}_d^- \end{pmatrix}, \quad \hat{S}. \tag{2}$$

In the $\mathbb{Z}_3$ implementation that we will assume, the Higgs superpotential is made up of two operators, each one introduces two dimensionless couplings λ and κ,

$$W_{Higgs} = -\lambda \hat{S}\hat{H}_d \cdot \hat{H}_u + \frac{1}{3}\kappa \hat{S}^3, \tag{3}$$

where $\hat{H}_d \cdot \hat{H}_u = \epsilon_{ab}\hat{H}_d^a \hat{H}_u^b$ and ϵ_{ab} is the two dimensional Levi-Civita symbol with $\epsilon_{12} = 1$. The two parameters λ and κ of the superpotential will, by construction, affect the phenomenology of both the Higgs and chargino/neutralino sector. λ in particular is crucial, not only it is necessary to induce the μ term but it also gives rise to mixing in the neutralino sector as well as in the Higgs sector between the Higgs doublets and the new singlet. For the purpose of parameter-counting and of the renormalisation of the Higgs sector at one-loop there is no need to go over the Yukawa superpotential which can be found in [22]. However, we do need to clearly specify again the soft SUSY breaking Lagrangian, in particular the part relating to the Higgs sector,

$$\begin{aligned} -\mathcal{L}_{soft,scalar} = &\, m_{H_u}^2 |H_u|^2 + m_{H_d}^2 |H_d|^2 + m_S^2 |S|^2 \\ &+ (\lambda A_\lambda H_u \cdot H_d S + \frac{1}{3}\kappa A_\kappa S^3 + \text{h.c.}) \end{aligned} \tag{4}$$

The first two terms in the first line represent the soft mass terms for the Higgs doublets and the third, not present in the MSSM, of the singlet. The second line, not present in the MSSM, represents the NMSSM trilinear Higgs couplings A_κ, A_λ. A_λ affects the mixing between the Higgs doublets and the singlet, beside the mixing introduced by λ. This paramater plays an important role in the phenomenology of the Higgs sector in the NMSSM, note that it gives rise to a Higgs tri-linear coupling $H_u H_d S$. No source of CP violation is assumed.

We are now in a position to write the Higgs potential whose parameters will need to be renormalised. With g, g' being respectively the $SU(2)$ and $U(1)$ gauge couplings and specifying the components of the doublets,

$$H_d = \begin{pmatrix} H_d^0 \\ H_d^- \end{pmatrix} \quad H_u = \begin{pmatrix} H_u^+ \\ H_u^0 \end{pmatrix}$$

the potential writes

$$\begin{aligned} V_{Higgs} = &|\lambda(H_u^+ H_d^- - H_u^0 H_d^0) + \kappa S^2|^2 + (m_{H_u}^2 + |\lambda S|^2)\left(|H_u^0|^2 + |H_u^+|^2\right) \\ &+ (m_{H_d}^2 + |\lambda S|^2)\left(|H_d^0|^2 + |H_d^+|^2\right) + \frac{g^2 + g'^2}{8}\left(|H_u^0|^2 + |H_u^+|^2 - |H_d^0|^2 - |H_d^-|^2\right)^2 \\ &+ \frac{g^2}{2}|H_u^+ H_d^{0*} + H_u^0 H_d^{-*}|^2 + m_S^2|S|^2 + (\lambda A_\lambda (H_u^+ H_d^- - H_u^0 H_d^0)S + \frac{1}{3}\kappa A_\kappa S^3 + \text{h.c.}). \end{aligned} \tag{5}$$

The physical scalar fields consist of 3 neutral CP-even Higgs bosons, h_1^0, h_2^0, h_3^0, 2 CP-odd Higgs bosons, A_1^0, A_2^0 and a charged Higgs boson, $H^\pm$. The NMSSM contains of course the SM. In particular the SM gauge parameters g, g' and $v^2 = v_u^2 + v_d^2$ are traded for the input parameters

IOP Conf. Series: Journal of Physics: Conf. Series **920** (2017) 012011 doi:10.1088/1742-6596/920/1/012011

e, M_W, M_Z. From the Higgs potential, equation (5), it is clear that the Higgs sector of the NMSSM depends on the parameters:

$$\underbrace{t_\beta, \lambda, \kappa, \mu}_{\text{in } \tilde{\chi} \text{ sector also}}, A_\lambda, A_\kappa, m_{H_d}, m_{H_u}, m_S. \tag{6}$$

The first four are also involved in the characterisation of the neutralino/chargino sector. Alternatively the last three soft Higgs masses can be traded for the tadpoles of the neutral Higgs which need to be constrained to zero to impose that the potential is at its minimum such that the independent parameters can be taken to be

$$\underbrace{t_\beta, \lambda, \kappa(m_\kappa), \mu}_{\text{in } \tilde{\chi} \text{ sector also}}, A_\lambda, A_\kappa, (\mathcal{T}_{H_d}, \mathcal{T}_{H_u}, \mathcal{T}_S). \tag{7}$$

The latter are therefore considered as physical observables. The other six parameters are not unambiguously defined in a simple mapping to an observable. We will have a lot to say about the choice and definitions of the input parameters that will construct the set of these six parameters. This issue is directly related to the renormalisation scheme.

3. Renormalisation

In a nutshell, renormalisation consists in adding (subtracting) the ultra-violet infinite parameters and introducing wave function renormalisation for all fields. Then we need to impose conditions that amount to a definition of the parameters. The procedure consists in the following steps

- From $\mathcal{G}_p = \underbrace{t_\beta, \lambda, \kappa, \mu}_{\text{in } \tilde{\chi} \text{ sector also}}, A_\lambda, A_\kappa, \left(M^2_{H_u}, M^2_{H_d}, M^2_S \to t_{H_d}, t_{H_u}, t_S\right), M_1, M_2 (\text{and } \underbrace{g, g', v \to e, M_W, M_Z}_{\text{SM}})$
  shift all (independent basic Lagrangian) parameters: $\mathcal{G}_p \to \mathcal{G}_p + \delta\mathcal{G}_p$
- this means that mass mixing will appear: non diagonal transition $A_i^0 Z^0, A_i^0 G^0, h_1 h_2, ...$ $(\delta m_{h_i h_j}, \delta m^2_{H^\pm G^\pm})$ are generated and diagonal masses are shifted $(\delta m^2_{h_i})$.
- No need to apply shifts to the diagonalising matrices $(S_h, U(\beta), \cdots)$, these are renormalised (no shift), same with gauge-fixing (not physical)

$$\begin{pmatrix} G^+ \\ H^+ \end{pmatrix}_0 = U(\beta) \begin{pmatrix} h_d^+ \\ h_u^+ \end{pmatrix}_0 \quad \text{implies also} \quad \begin{pmatrix} G^+ \\ H^+ \end{pmatrix} = U(\beta) \begin{pmatrix} h_d^+ \\ h_u^+ \end{pmatrix}.$$

- In any case field renormalisation (before or after rotation) still needed

$$\begin{pmatrix} h_1^0 \\ h_2^0 \\ h_3^0 \end{pmatrix}_0 = Z_S \begin{pmatrix} h_1^0 \\ h_2^0 \\ h_3^0 \end{pmatrix}, \begin{pmatrix} A_1^0 \\ A_2^0 \\ G^0 \end{pmatrix}_0 = Z_P \begin{pmatrix} A_1^0 \\ A_2^0 \\ G^0 \end{pmatrix}, \begin{pmatrix} G^\pm \\ H^\pm \end{pmatrix}_0 = Z_C \begin{pmatrix} G^\pm \\ H^\pm \end{pmatrix} ((Z_S)_{ij} = 1_{ij} + \delta Z_{h_i h_j}/2)$$

- As an example, the renormalised self-energies of the CP-even Higgs scalars can be cast in the form

$$\left\{ \begin{array}{lcl} \hat{\Sigma}_{h_i^0 h_i^0}(p^2) & = & \Sigma_{h_i^0 h_i^0}(p^2) - \delta m^2_{h_i^0} + (p^2 - m^2_{h_i^0})\delta Z_{h_i^0} \\ \hat{\Sigma}_{h_i^0 h_j^0}(p^2) & = & \Sigma_{h_i^0 h_j^0}(p^2) - \delta m^2_{h_i^0 h_j^0} + \frac{1}{2}(p^2 - m^2_{h_i^0})\delta Z_{h_i^0 h_j^0} + \frac{1}{2}(p^2 - m^2_{h_j^0})\delta Z_{h_j^0 h_i^0} \end{array} \right. \tag{8}$$

We then require that *i)* Mixing vanishes between physical states when these are on-shell, *essentially* (to solve for δZ_{ij}'s); *ii)* the residue at the pole (mass) of the propagator is 1. The other conditions are set by using/choosing a minimum/sufficient set of physical masses as input

parameters except α_{em} which defined like in QED from the Thomson limit. In this approach only two point-functions are needed. The big question is which minimum set of parameters to choose?
Any choice will require solving a coupled system of counterterms, in other words setting the passage from the underlying Lagrangian parameters to, if possible, physical parameters such as masses. Leaving aside the SM parameter,

$$\begin{pmatrix} \delta\texttt{input}_1 \\ \cdots \\ \cdots \\ \delta\texttt{input}_8 \end{pmatrix}_{\chi^\pm,\chi_i^0,h_i^0,A_i^0,H^\pm} = \mathcal{P}_{8,\text{param.}} \begin{pmatrix} \delta M_1 \\ \delta M_2 \\ \delta\kappa \\ \delta\mu \\ \delta\lambda \\ \delta t_\beta \\ \delta A_\lambda \\ \delta A_\kappa \end{pmatrix} + \mathcal{R}_{n,\text{residual}}, \tag{9}$$

$\mathcal{R}_{8,\text{residual}}$ counterterms such as gauge couplings, etc (SM). Inverting the system means that the corresponding determinant($\rightarrow 1/\text{Det}(\mathcal{P}_{m,\text{param.}})$) better be not too small otherwise the finite part of the counterterms could introduce unnecessarily large radiative *corrections (sic!)*. It is best to break up the system, into smaller subsystem $\mathcal{P}_{n,\text{param.}} = \mathcal{P}_{m,\text{param.}} \oplus \mathcal{P}_{p,\text{param.}} \oplus \cdots, \quad m + p + \cdots = n$. At each step one should avoid a choice such that $\text{Det}(\mathcal{P}_{m,\text{param.}}) \rightarrow 0$. For example picking up a wino-like neutralino to reconstruct M_1 would be a bad choice if the mixing between the bino and the wino is small. Since t_β is central, the easiest set-up is to define t_β in $\overline{\text{DR}}$ with t_β extracted independently from the Higgs sector (through wave function renormalisation condition) $\delta t_\beta/t_\beta = \left[\frac{1}{2}(\delta Z_{H_u} - \delta Z_{H_d})\right]_\infty$, "$\infty$ means we only keep the infinite part. In this mixed scheme, μ, M_2 are easily constrained from the charginos masses. Note however that though algebraically this conversion is straighforward, it is not sure that the assignation of μ, M_2 to the correct chargino mass will be correct. Again a knowledge of which is which is essential. One needs more than the masses. We are then left with having to define $M_1, \kappa, (\lambda)$ from neutralino masses again with the caveat that we somehow know the composition. This may not be a big hurdle, since if the masses are to be measured, this means they have been produced. If we know in which proportion this could give a valuable hint on the nature of the particle. $A_\lambda, A_\kappa(\lambda)$ can only be extracted from the Higgs (A, h, H^+). If this were possible with t_β from $\overline{\text{DR}}$ then the break-up corresponds to

$$\mathcal{P}_8 = \mathcal{P}_{1,t_\beta} \oplus \mathcal{P}_{2,\chi_{1,2}^\pm} \oplus \mathcal{P}_{3,\chi^0} \oplus \mathcal{P}_{2,A_1^0,A_2^0} \text{ or } \mathcal{P}_8 = \mathcal{P}_{1,t_\beta} \oplus \mathcal{P}_{2,\chi_{1,2}^\pm} \oplus \mathcal{P}_{2,\chi^0} \oplus \mathcal{P}_{3,H^\pm,A_1^0,A_2^0(h^0)}$$

A complete OS scheme, based on the input provided by 8 masses would look like $\mathcal{P}_{8;\chi_{1,2}^\pm,\chi_{1,2,3}^0,H^\pm,A_1^0,h^0}$. We have used different variants which nonetheless all take the chargino masses as input. *i*)$\text{OS}_{ijkA_1A_2H^+}$ with the masses of 3 χ^0 preferably $\tilde{b}, \tilde{h}, \tilde{s}$-dominated; *ii*)$\text{OS}_{ijh_\alpha A_1A_2H^+}$ (only 2 neutralinos) and *iii*)$\text{OS}_{ih_\alpha h_\beta A_1A_2H^+}$ (only one neutralino,$\tilde{b}$).

In renormalisable theories, most loop calculations involving 1-,2- and 3-point functions give rise to ultraviolet infinities. The latter require a regularisation. In dimensional reduction/regularisation these infinities are contained in the quantity $\widetilde{C}_{UV} = 2/\epsilon - \gamma_E + \ln(4\pi/\bar{\mu}^2) = 2/\epsilon - \gamma_E + \ln(4\pi) + \ln(1/\bar{\mu}^2)$ where the continuation to $n = 4 - \epsilon$ dimension require the introduction of the compensating regularisation scale $\bar{\mu}$. The definition of an underlying parameter p_i at one-loop, say OS scheme, based on a physical observable calls for the corresponding counterterm $\delta p_i/p_i = \beta_{p_i}(C_{UV} + \ln(Q_{p_i}/\bar{\mu}))$ $(\beta_{p_i} = \partial p_i/\partial \ln 1/\bar{\mu})$. Q_{p_i} is a scheme-dependent an effective scale, which depends on the point where the parameter has been

IOP Conf. Series: Journal of Physics: Conf. Series **920** (2017) 012011 doi:10.1088/1742-6596/920/1/012011

defined and other masses that may appear at one-loop in the definition of the parameter. β_{p_i} is a universal, scheme independent factor. To track down the influence of the scheme on the correction to an observable, it is useful to know the parametric dependence of the observable on the parameter p_i's, $\partial O/\partial p_i = \kappa_{p_i}$. In our so-called $\overline{\rm DR}$ scheme we will only keep $\beta_{p_i} C_{UV}$, the "finite part" is set to 0. In On-shell (OS) scheme the "finite part" is $\beta_{p_i} \ln(Q_{p_i}/\bar{\mu})$. Then the full one-loop correction to the observable $\delta\mathcal{O}/\mathcal{O} = \overbrace{\Delta(C_{UV} + \ln(Q_\Delta/\bar{\mu}))}^{virtual} + \overbrace{\sum_i \kappa_i \delta p_i/p_i}^{Counterterms}$ leads to $\delta\mathcal{O}^{\rm OS}/\mathcal{O} = \sum_i \beta_{p_i}\kappa_{p_i} \ln(Q_{p_i}/Q_\Delta)$ $\quad \delta\mathcal{O}^{\rm DR}/\mathcal{O} = \ln(\bar{\mu}/Q_\Delta)\sum_i \beta_{p_i}\kappa_{p_i})$. A good choice of $\bar{\mu}$ or Q_{p_i} will minimize the correction. These simple observations show that one can have large corrections when/if $i)\beta_p$ is large, and/or $ii)\kappa_p$ is large and/or $iii)$ when there is a large mismatch/difference between the scales Q_Δ and Q_p.

4. Applications

We have considered two scenarios. We will not, in this short write-up, give the full details of both models by listing all their defining underlying parameters. The salient differences is that point A is MSSM-like with a small $\lambda = 0.1$ while point B has $\lambda = 0.67$. In order for A to reproduce a standard model-like Higgs with mass 125GeV, the SUSY scale set by the masses of the stops as well as, more crucially, A_t need to be large. A large t_β helps also. For example for point A, $A_t/A_\lambda \sim 27$ while for point B $A_t/A_\lambda \sim 2.5$. While a large A_t helps obtain a large correction for the Higgs mass it also provides a very large β function for A_λ. The latter is induced by an enhanced top Yukawa coupling modulated by A_t/A_λ as be seen from the RGE's. With h_t the Yukawa coupling, the dominant contributions to the running of the underlying couplings of the NMSSM can be cast as[24]

$$\frac{1}{h_t^2}\frac{dh_t^2}{d\tau} = \frac{1}{A_t}\frac{dA_t}{d\tau} = \frac{2}{\lambda^2}\frac{d\lambda^2}{d\tau} = \frac{2}{\mu^2}\frac{d\mu^2}{d\tau} = 6h_t^2 \quad \text{and} \quad \frac{1}{A_\lambda}\frac{dA_\lambda}{d\tau} = 3h_t^2\frac{A_t}{A_\lambda}, \tag{10}$$

with $\tau = \ln\bar{\mu}^2/16\pi^2$. Numerically and with no approximation we find for the beta functions expressed in units of 10^{-3},
Point A: $\beta_\mu = -11.40, \beta_{t_\beta} = 16.9, \beta_\lambda = -11.65, \beta_\kappa = -0.76, \beta_{A_\lambda} = -1097.4$ and
Point B: $\beta_\mu = -14.25, \beta_{t_\beta} = 17.63, \beta_\lambda = -20.45, \beta_\kappa = -18.57, \beta_{A_\lambda} = -122.7$
It is also useful to list the finite parts of the counterterms in some of the schemes for both points A and B.

Point A The finite parts computed at $\bar{\mu} = \mathrm{Q_{susy}} = 1117.25\mathrm{GeV}$ give

$$(\delta\mu/\mu, \delta t_\beta/t_\beta, \delta\lambda/\lambda)_{\rm finite} = \overbrace{(-2.42\%, 0, 62.26\%)}^{t_{134A_1A_2}}; \overbrace{(-1.57\%, -80.69\%, -7.88\%)}^{{\rm OS}_{34h_2A_1A_2H^+}} \tag{11}$$

$$(\delta\kappa/\kappa, \delta A_\lambda/A_\lambda, \delta A_\kappa)_{\rm finite} = \overbrace{(64.01\%, -5.49\%, 0.65)}^{t_{134A_1A_2}}; \overbrace{(-6.01\%, 134\%, 0.66)}^{OS_{34h_2A_1A_2H^+})}. \tag{12}$$

Point B Here the finite parts" computed at $\bar{\mu} = \mathrm{Q_{susy}} = 753.55\mathrm{GeV}$

$$(\delta\mu/\mu, \delta t_\beta/t_\beta, \delta\lambda/\lambda)_{\rm finite} = \overbrace{(-1.04\%, 0, 3.71\%)}^{t_{123A_1A_2}}; \overbrace{(-1.63\%, 6.49\%, 5.94\%)}^{{\rm OS}_{12h_2A_1A_2H^+}} \tag{13}$$

$$(\delta\kappa/\kappa, \delta A_\lambda/A_\lambda, \delta A_\kappa)_{\rm finite} = \overbrace{(3.25\%, 6.85\%, 10.84)}^{t_{123A_1A_2}}; \overbrace{(6.05\%, 3.40\%, 11.54)}^{OS_{12h_2A_1A_2H^+}}. \tag{14}$$

IOP Conf. Series: Journal of Physics: Conf. Series **920** (2017) 012011 doi:10.1088/1742-6596/920/1/012011

All the schemes t where t_β is defined as $\overline{\rm DR}$ and the other parameters are taken as OS take the chargino masses as inputs. The other parameters are explicit if one note that $1, 2, 3$ stand for the neutralinos ordered by increasing mass. The β functions as well as the finite parts of the counterterms give an insight on the value of the full one-loop corrections and more importantly on the differences between the results of the schemes. The parametric dependence is weighed by perturbing slightly around the values of the model. Fig. 2 shows that for Point A the dependence

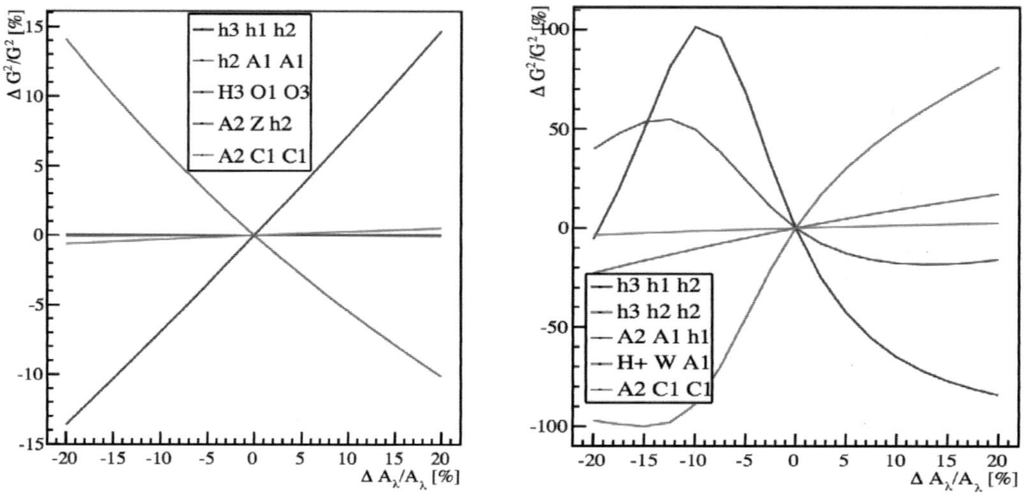

Figure 2. Dependence on A_λ of the square of the couplings, G_{ijk}, shown in the insert of the figures for Point A (left panel) and Point B (right panel). G_{ijk} represents the coupling constant of the interaction between ijk. The variations shown do not take into account the changes in the decay due to phase space that can also change because some of the masses involved in the decay change when the underlying parameters change. h_i stand for the CP-even Higgses. O are for the neutralinos and C the charginos.

one some of the underlying parameters for some of the chosen decays are practically flat or linear in contrast to Point B where the dependence is quite strong. For point B a small change in A_λ can trigger a large difference with a highly non linear variation.

We have computed the full one-loop electroweak corrections to various Higgs decays either to other Higgses or to Higgs gauge bosons or to a pair of neutralinos/charginos. Table 1 is for Point A. To understand the scheme dependence of the majority of the most important decays of the different Higgses, it is important to keep in mind the singlet nature of the states involved. Recall that this scenario is characterised by very small mixing whereby the Higgs states h_2^0, A_1^0 and the neutralino χ_3^0 are essentially singlet states. There are three classes of decays, those among the predominantly singlets ($h_2^0 \to A_1^0 A_1^0$), those not involving any of these states($A_2^0 \to \tilde{\chi}_1^+\tilde{\chi}_1^-, h_3^0 \to \tilde{\chi}_1^+\tilde{\chi}_1^-$ and $A_2^0 \to \tilde{\chi}_1^0\tilde{\chi}_1^0$) with characteristics close to the MSSM and with very little dependence on λ. The last class involves both singlet and non singlet states. The latter decays are therefore sensitive to the mixing parameters due to the addition of the singlet in the NMSSM. For most decays this mixing parameter is essentially λ while for $h_3h_2h_1$ A_λ is crucial. These observations together with the numerical values of the β functions and the values of the finite terms of the counterterms can explain the differences. The mixed scheme $t_{134A_1A_2|}$ performs quite badly for almost all decays that involve singlet/non singlet states. The latter are very sensitive to λ which is badly reconstructed in this scheme. This also explains the almost equal value of the correction in the many channels, $\sim$ 120% which is mainly accounted for by

Table 1. Point A. Corrections to the partial decay widths of Higgs bosons in other Higgs bosons and/or neutralinos, charginos and gauge bosons. The column TL represents the tree-level (in units of GeV/100) partial width. The other columns give the percentage relative full one-loop correction, in the mixed scheme where only t_β is taken $\overline{\rm DR}$ ($t_{134A_1A_2}|$), the full OS scheme and full $\overline{\rm DR}$. For $t_{134A_1A_2}|$ and $\overline{\rm DR}|$ the scale $\bar{\mu}$ is taken at the mass of the decaying Higgs, for $\overline{\rm DR}\,Q_{\rm SUSY}$ the scale at the SUSY scale.

Decays	**TL** $\times 10^{-2}$	$t_{134A_1A_2}\|$	$\rm OS_{34h_2A_1A_2H^+}$	$\overline{\rm DR}\|$	$\overline{\rm DR}\,\|Q_{\rm SUSY}$
$h_2^0 \to A_1^0 A_1^0$	4.79	128%	-12%	0.4%	-0.4%
$h_3^0 \to h_1^0 h_2^0$	2.21	116%	79%	52%	-1.7%
$h_3^0 \to \tilde{\chi}_1^0 \tilde{\chi}_3^0$	3.52	122%	-3%	2%	0.3%
$h_3^0 \to \tilde{\chi}_2^0 \tilde{\chi}_3^0$	3.38	126%	-35%	3%	1.1%
$h_3^0 \to \tilde{\chi}_1^+ \tilde{\chi}_1^-$	4.55	1%	-11%	-9%	-7.4%
$A_2^0 \to Z h_2^0$	1.86	120%	80%	-56%	-14.5%
$A_2^0 \to \tilde{\chi}_1^0 \tilde{\chi}_1^0$	3.30	28%	13%	0.3%	-1.6%
$A_2^0 \to \tilde{\chi}_1^0 \tilde{\chi}_3^0$	2.44	130%	-31%	8%	6.2%
$A_2^0 \to \tilde{\chi}_2^0 \tilde{\chi}_3^0$	3.02	122%	-5%	-0.4%	-1.9%
$A_2^0 \to \tilde{\chi}_1^+ \tilde{\chi}_1^-$	5.51	-10%	-1.5%	-6%	-8%
$H^+ \to W^+ h_2^0$	2.01	119%	79%	-56%	-16%
$H^+ \to \tilde{\chi}_1^+ \tilde{\chi}_3^0$	6.40	125%	-18%	3%	1.1%

the finite part of $2\delta\lambda$ in this scheme. The decay $h_3^0 \to h_1^0 h_2^0$ (and to some extent $A_2^0 \to Z h_2^0$) stands apart. It is the only channel where there is a large running, as testified by the comparison between the two $\overline{\rm DR}$ schemes. In fact, this coupling depends on A_λ whose running is quite well approximated by the large value of β_{A_λ}.

For point B, the value of λ is large enough, this is reflected also in the fact the partial widths are for all channels an order of magnitude larger than for point A. A discussion in terms of almost pure or singlet states is not appropriate. For this more genuine NMSSM, the OS scheme performs quite well. It rests that the scheme and scale dependence for $h_3^0 \to h_1^0 h_2^0$ is quite large. This is due not so much to β_{A_λ} bur rather to the strong parametric dependence on A_λ as seen in Fig. 2.

5. Summary and conclusions

Full renormalisation (all sectors) of the NMSSM at one-loop is now completed and implemented in `SloopS`. The set-up allows to choose between different on-shell schemes and also "mixed schemes" with one or several parameters through a $\overline{\rm DR}$ subtraction. The on-shell schemes are at the moment all based on the usage of physical masses and hence the use of two-point functions in the definition of the parameters. Especially in scenarios where the singlet mixing parameter is small, these schemes are not optimal in extracting this crucial mixing parameter nor in extracting t_β. We must go beyond taking only masses as inputs. When new particles are discovered, not only their masses will be measured but the way they are produced, the strengths of their production and decays offer an important handle that may not need the reconstruction of the whole spectrum. This is technically much more challenging, but it is possible (at least in some manifestations). In the MSSM $A^0 \to \tau\bar{\tau}$ was shown to be an excellent input for t_β[19].

I would like to end with a personal note. Last time I saw Shimizu-sensei was in February 2015. The last dinner I had with him was in a Fugu restaurant. As always with him, *un*

Table 2. Point B. Corrections to the partial decay widths of Higgs bosons in other Higgs bosons and/or neutralinos, charginos and gauge bosons. TL represents the partial decay widths (in GeV). The percentage relative full one-loop corrections are given as percentage relative corrections in the OS scheme, namely, $\mathrm{OS}_{12h_2A_1A_2H^+}$ as well as a full $\overline{\mathrm{DR}}$ at the SUSY scale and the mass of the decaying Higgs.

	TL	OS	$\overline{\mathrm{DR}}\vert$	$\overline{\mathrm{DR}}\, Q_{\mathrm{SUSY}}$
$h_3^0 \to \tilde{\chi}_1^0\tilde{\chi}_2^0$	0.727	14%	5%	3%
$h_3^0 \to A_1^0 Z$	0.613	3%	-3%	-8 %
$h_3^0 \to h_2^0 h_1^0$	0.341	-25%	-106%	-50%
$h_3^0 \to h_2^0 h_2^0$	0.514	6%	13%	-28%
$A_2^0 \to \tilde{\chi}_1^+\tilde{\chi}_1^-$	1.523	7%	2%	1%
$A_2^0 \to \tilde{\chi}_1^0\tilde{\chi}_1^0$	0.723	32%	2%	2%
$A_2^0 \to Z h_2^0$	0.638	12%	-16%	-9%
$A_2^0 \to A_1^0 h_1^0$	0.415	-0.3%	-32%	-17%
$H^+ \to \tilde{\chi}_1^+\tilde{\chi}_2^0$	1.056	6%	10%	8%
$H^+ \to W^+ h_2^0$	0.609	11%	-18%	-10%
$H^+ \to W^+ A_1^0$	0.603	2%	-3%	-9%
$H^+ \to \tilde{\chi}_1^+\tilde{\chi}_1^0$	0.561	21%	9%	9%

fin gourmet, the meal was excellent as were all the subtleties about the dishes. The next morning, in a email, he told me he enjoyed having dinner with me (and with Kato-san and Kuroda-san) but he thought it was the last time we saw each other. I did not believe it, but he was right. Back to France, we exchanged regularly. Most of exchanges were about food and pottery. His celadon gifts to me are priceless. We did discuss physics and of all the exchanges the one which moved me the most was about the visit of Nakawaza-san to the hospital when we told me again about the 2-loop electroweak calculations project. So much about my 2015 trip to Japan.

The last trip of Shimizu to France, Paris and Annecy, was wonderful for my family and myself. I was told by a dear common friend that his dream and one of his wishes had been to take his dearest friends to see Paris. These friends are always welcome to France and I very much hope you will visit. And when you come we will go to `Minami`...in Annecy! Minami is a new, small, genuine Japanese restaurant that opened about 4 or 5 years ago. How wonderful that I can get back to the days of Minami-Tateya by walking 2 minutes from my home and step in *Minami* like Shimizu did in January 2013.

Figure 3. Shimizu-sensei in front of the Japanese restaurant in Annecy, the Minami! January 2013.

Acknowledgments
I first of all would like to thank Shimizu-sensei for the intense scientific collaboration I enjoyed with him and the Minami-Tateya Group. On a personal level he has helped me go through some very difficult times and shared with me some wonderful moments, not to mention introducing me to so many subtleties of Japanese culture. I would also like to thank all those who have been by his side during the last months, days and hours and took much of their time and energy to comfort him and who were kind enough to report to me from time to time. I also thank the large circle of Minami-Tateya for making this gathering, the trip to Hakone and the rest place of Shimizu-sensei possible and enjoyable. I can not be thankful enough for allowing me to meet some of Shimizu's friends I had not known before and with whom I promise to keep in touch! Thank you!

References

[1] Dombey N and Boudjema F 1990 *NATO Sci. Ser. B* **233** pp.1–578
[2] Fujimoto J, Igarashi M and Shimizu Y 1987 *Prog. Theor. Phys.* **77** 118
[3] Boudjema F *et al.* 1996 Standard model processes *Workshop on Physics at LEP2, v.1, G. Altarelli (ed.), S. Torbjrn (ed.) and F. Zwirner (ed.)* pp 207–248 [207(1996)] (*Preprint* hep-ph/9601224) URL https://inspirehep.net/record/415134/files/arXiv:hep-ph_9601224.pdf
[4] Boudjema F and Chopin E 1996 *Z. Phys.* **C73** 85–110 (*Preprint* hep-ph/9507396)
[5] Belanger G, Boudjema F, Fujimoto J, Ishikawa T, Kaneko T, Kato K, Lafage V, Nakazawa N and Shimizu Y 1999 Implementation of the nonlinear gauge into GRACE *6th International Workshop on New Computing Techniques in Physics Research: Software Engineering, Artificial Intelligence Neural Nets, Genetic Algorithms, Symbolic Algebra, Automatic Calculation (AIHENP 99) Heraklion, Crete, Greece, April 12-16, 1999* (*Preprint* hep-ph/9907406) URL http://alice.cern.ch/format/showfull?sysnb=0320104
[6] Belanger G, Boudjema F, Fujimoto J, Ishikawa T, Kaneko T, Kato K and Shimizu Y 2003 *Nucl. Phys. Proc. Suppl.* **116** 353–357 [,353(2002)] (*Preprint* hep-ph/0211268)
[7] Belanger G, Boudjema F, Fujimoto J, Ishikawa T, Kaneko T, Kato K and Shimizu Y 2003 *Phys. Lett.* **B559** 252–262 (*Preprint* hep-ph/0212261)
[8] Belanger G, Boudjema F, Fujimoto J, Ishikawa T, Kaneko T, Kato K, Shimizu Y and Yasui Y 2003 *Phys. Lett.* **B571** 163–172 (*Preprint* hep-ph/0307029)
[9] Belanger G, Boudjema F, Fujimoto J, Ishikawa T, Kaneko T, Kurihara Y, Kato K and Shimizu Y 2003 *Phys. Lett.* **B576** 152–164 (*Preprint* hep-ph/0309010)
[10] Boudjema F, Fujimoto J, Ishikawa T, Kaneko T, Kato K, Kurihara Y, Shimizu Y, Yamashita S and Yasui Y 2004 *Nucl. Instrum. Meth.* **A534** 334–338 (*Preprint* hep-ph/0404098)
[11] Boudjema F, Fujimoto J, Ishikawa T, Kaneko T, Kato K, Kurihara Y, Shimizu Y and Yasui Y 2004 *Phys. Lett.* **B600** 65–76 (*Preprint* hep-ph/0407065)
[12] Boudjema F, Fujimoto J, Ishikawa T, Kaneko T, Kato K, Kurihara Y and Shimizu Y 2004 *Nucl. Phys. Proc. Suppl.* **135** 323–327 [,323(2004)] (*Preprint* hep-ph/0407079)
[13] Boudjema F, Fujimoto J, Ishikawa T, Kaneko T, Kurihara Y, Shimizu Y, Kato K and Yasui Y 2005 *eConf* **C050318** 0601 (*Preprint* hep-ph/0510184)
[14] Kato K, Boudjema F, Fujimoto J, Ishikawa T, Kaneko T, Kurihara Y, Shimizu Y and Yasui Y 2006 *PoS* **HEP2005** 312
[15] Belanger G, Boudjema F, Fujimoto J, Ishikawa T, Kaneko T, Kato K and Shimizu Y 2006 *Phys. Rept.* **430** 117–209 (*Preprint* hep-ph/0308080)
[16] Semenov A 2009 *Comput. Phys. Commun.* **180** 431–454 (*Preprint* 0805.0555)
[17] Hahn T 2000 *Nucl. Phys. Proc. Suppl.* **89** 231–236 (*Preprint* hep-ph/0005029)
[18] Baro N, Boudjema F and Semenov A 2008 *Phys. Lett.* **B660** 550–560 (*Preprint* 0710.1821)
[19] Baro N, Boudjema F and Semenov A 2008 *Phys. Rev.* **D78** 115003 (*Preprint* 0807.4668)
[20] Baro N and Boudjema F 2009 *Phys.Rev.* **D80** 076010 (*Preprint* 0906.1665)
[21] Baro N, Boudjema F, Chalons G and Hao S 2010 *Phys. Rev.* **D81** 015005 (*Preprint* 0910.3293)
[22] Belanger G, Bizouard V, Boudjema F and Chalons G 2016 *Phys. Rev.* **D93** 115031 (*Preprint* 1602.05495)
[23] Bizouard V 2015 *Precision calculations in the Next-to-Minimal Supersymmetric Standard Model* Theses Université Grenoble Alpes (*Preprint* https://hal.archives-ouvertes.fr/tel-01447488)
[24] Ellwanger U, Hugonie C and Teixeira A M 2010 *Phys. Rept.* **496** 1–77 (*Preprint* 0910.1785)

IOP Conf. Series: Journal of Physics: Conf. Series **920** (2017) 012012 doi:10.1088/1742-6596/920/1/012012

ISR effects on loop corrections of a top pair-production at the ILC

Nhi M. U. Quach

The Graduate University for Advanced Studies (SOKENDAI), Hayama, Kanagawa 240-0193, Japan, and
High Energy Accelerator Research Organization (KEK), Tsukuba, Ibaraki 305-0801, Japan.

E-mail: nhiquach@post.kek.jp

Yoshimasa Kurihara

High Energy Accelerator Research Organization (KEK), Tsukuba, Ibaraki 305-0801, Japan.

E-mail: yoshimasa.kurihara@kek.jp

Abstract. Precise predictions for an $e^+e^- \to t\bar{t}$ cross section are presented at an energy region from 400 GeV to 800 GeV. Cross sections are estimated including the beam-polarization effects with full $\mathcal{O}(\alpha)$, and also with effects of the initial-state photon emission. A resummation technique is used for the initial-state photon emission up to two-loop order. A pure-weak correction is defined as the full electroweak corrections without the initial-state photonic corrections. As a result, it is obtained that the total cross section of a top quark pair-production receives the pure-weak corrections of +4% over the trivial initial state corrections at a center of mass energy of 500 GeV. Among the initial state contributions, a contribution from two-loop diagrams gives −5% correction over the one-loop ones at the center of mass energy of 500 GeV.

1. Introduction

The standard theory of particle physics are established finally by a discovery of the Higgs boson[1, 2] in 2012. A current target of particle physics is searching for a more fundamental theory beyond the standard theory (BST). A keystone along this direction must be the Higgs boson and top quark. Since the top quark is the heaviest fermion with a mass at the electroweak symmetry-breaking scale, it is naturally expected to have a special role in the BST.

The international linear collider[3] (ILC), which is an electron-positron colliding experiment with centre of mass (CM) energies above 250 GeV, is proposed and intensively discussed as a future project of high-energy physics. One of main goals of ILC experiments is a precise measurement of top quark properties. Detailed Monte Carlo studies have shown that the ILC would be able to measure most of the standard model parameters to within sub-percent levels[4]. Because of the improvement of the experimental accuracy of the ILC, theoretical predictions are required be given with new level of precision. In particular, a radiative correction due to the electroweak interaction (including spin polarizations) is mandatory for such requirements. Before the discovery of the top quark, a full electroweak radiative correction was conducted for an $e^-e^+ \to t\bar{t}$ process at a lower energy[5], and was then obtained independently for higher energies[6, 7]. The same correction fo a $e^-e^+ \to t\bar{t}\gamma$ process has also been reported[8]. Recently, full electroweak radiative corrections for the process $e^-e^+ \to t\bar{t} \to b\bar{b}\mu^+\mu^-\nu_\mu\bar{\nu}_\mu$ using a narrow-width approximation for the top quarks including the spin-polarization effects are reported[9] by authors including those of a present report. In this report, the precise estimation of the initial-state photon-radiations is discussed in detail

2. Calculation Method

2.1. GRACE system

For precise cross-section calculations of target process, a GRACE-Loop system is used in this study. The GRACE system is an automatic system to calculate cross sections of scattering processes at one-loop level for the standard theory[10] and the minimum SUSY model[11]. The GRACE system has treated electroweak processes with two, three or four particles in the final state are calculated[12, 13, 14, 15]. The renormalization of the electroweak interaction is carried out using on-shell scheme[16, 17]. Infrared divergences are regulated using fictitious photon-mass[17]. The symbolic manipulation package FORM[18] is used to handle all Dirac and tensor algebra in n-dimensions. For loop integrations, all tensor one-loop integrals are reduced to scalar integrals using our own formalism, then performed integrations using packages FF[19] or LoopTools[20]. Phase-space integrations are done using an adaptive Monte Carlo integration package BASES[21, 22]. For numerical calculations, we use a quartic precision for floating variables.

In the GRACE system, while using R_ξ-gauge in linear gauge-fixing terms, the non-linear gauge fixing Lagrangian[23, 10] is employing for the sake of the system checking. Before calculating cross sections, we performed numerical tests to confirm that the amplitudes are independent of all redundant parameters around 20 digits at several randomly chosen phase points. In addition to above checks, soft-photon cut-off independence was examined: cross sections at the one-loop level, results must be independent from a head-photon cut-off parameter k_c. We confirmed that, while varying a parameter k_c from 10^{-4} GeV to 10^{-1} GeV, the results of numerical phase-space integrations are consistent each other within the statistical errors.

2.2. Structure function method

The effect of the initial photon emission can be factorized when a total energy of emitted photons are small enough compared with a beam energy. The calculations under a such approximation is referred as the "soft photon approximation(SPA)". Under the SPA, the corrected cross sections with the initial state photon radiation(ISR), σ_{ISR}, can be obtained from the tree cross sections σ_{Tree} using a structure function $H(x, s)$ as follows:

$$\sigma_{ISR} = \int_0^1 dx\ H(x,s)\sigma_{Tree}\left(s(1-x)\right), \tag{1}$$

where s is the CM energy square and x is a energy fraction of emitted photons. The structure function can be calculated using the perturbative method with the SPA. Concrete formulae of the structure function are known up to two loop order[17]. A further improvement of the cross section estimation is possible using the "exponentiation method". Initial state photon emissions under the SPA, a probability to emit each photon should be independent each other. Thus, the probability to emitt any number of photons can be calculated as

$$p = \sum_{i=0}^{\infty} p_i = \sum_{i=0}^{\infty} \frac{1}{i!}(p_1)^i, \tag{2}$$

where p_i is a probability to emit i-photons. A factor $1/i!$ appeared because of a number of identical particles (photons) in the final state. This is nothing more than the Taylor expansion of the exponential function. Therefore, the effect of the multiple photon emissions can be estimated by putting the one photon emission probability in an argument of the exponential function. This technique is referred as the exponentiation or the resummation method.

When the resummation method is applied to the cross section calculations at loop level, the corrected cross sections can not be expressed simply like the formula (1), because the same loop corrections are included in both of the structure function and the loop amplitudes. To avoid a double counting of the same effects both in the structure function and the loop amplitudes, one have to rearrange terms of corrections as follows: The total cross section at one-loop (fixed) order without the resummation, $\sigma_{NLO;fixed}$, can be expressed as

$$\sigma_{NLO;fixed} = \sigma_{Loop} + \sigma_{Hard} + \sigma_{Tree}\delta_{SPA} + \sigma_{Tree}, \tag{3}$$

where σ_{Loop} and σ_{Hard} are the cross sections from the loop and (hard) real-emission corrections, respectively. A factorized function δ_{SPA} is obtained form the real-radiation diagrams using the SPA. Here the hard photon is defined as the photon whose energy is greater than the threshold energy k_c. Final results must be independent of the value of k_c. The cross section with fixed order correction (3) can be improved using the resummation method as follows:

$$\sigma_{NLO;ISR} = \sigma_{Loop} + \sigma_{Hard} + \sigma_{Tree}\tilde{\delta}_{SPA} + \sigma_{ISR}, \tag{4}$$

$$\tilde{\delta}_{SPA} = \delta_{SPA} - \delta_{ISR}, \tag{5}$$

where σ_{ISR} shows the improved cross section using the structure function method (1) with the resummation, and δ_{ISR} is a correction factor from the initial state photon-loop diagrams, which is also included in the structure function $H(s, x)$.

3. Results and discussions

3.1. Input parameters

The input parameters used in this report are listed in Table 1. The masses of the light quarks (i.e., other than the top quark) and W boson are chosen to be consistent with low-energy experiments[24]. Other particle masses are taken from recent measurements[25]. The weak mixing angle is obtained using the on-shell condition $\sin^2\theta_W = 1 - m_W^2/m_Z^2$ because of our renormalization scheme. The fine-structure constant $\alpha = 1/137.0359859$ is taken from the low-energy limit of Thomson scattering, again because of our renormalization scheme.

Because the initial state photonic corrections are independent of the beam polarization, all cross section calculations in this report are performed with 100% left (right) polarization of an electron (positron), respectively.

3.2. Electroweak radiative corrections

At first, let us look at the results of the fixed order correction without a resummantion method shown in Fig. 1. The NLO calculations near the top-quark production threshold shows negative corrections about 7% at the CM energy of 400 GeV. This large correction is a so-called "coulomb correction", which is caused by that produced particles are moving slowly and have enough time duration to interact each other. The corrections becomes very small around the CM energy of 500 GeV and increases at the high energy region as to 2.8% at the CM energy of 800 GeV.

Among several types of the radiative correction, e.g., the initial and final state photon radiations, the vertex and box correction, and so on, one can see the initial state photonic corrections give a largest contribution at the high energy region. As shown in Fig. 2, the cross sections at tree level including the ISR correction (a gray line in the figure) are almost the same as the full order $\mathcal{O}(\alpha)$ electroweak correction (an orange line in the figure) at the CM energy above 700 GeV. This means that the main contribution of the higher order corrections is caused by the initial state photonic corrections. On the other hand, other corrections form the loop diagrams also give a large correction near the threshold region.

After subtracting a (rather trivial) ISR correction from the total corrections, one can discuss the "pure" weak correction in the full $\mathcal{O}(\alpha)$ electroweak radiative corrections. While the NLO correction degree is defined as

$$\delta_{NLO} = \frac{\sigma_{NLO;fixed} - \sigma_{Tree}}{\sigma_{Tree}}, \qquad (6)$$

u-quark mass	58.0×10^{-3} GeV	d-quark mass	58.0×10^{-3} GeV
c-quark mass	1.5 GeV	s-quark mass	92.0×10^{-3} GeV
t-quark mass	173.5 GeV	b-quark mass	4.7 GeV
Z-boson mass	91.187 GeV	W-boson mass	80.370 GeV
Higgs mass	126 GeV		

Table 1. Particle masses used in this report.

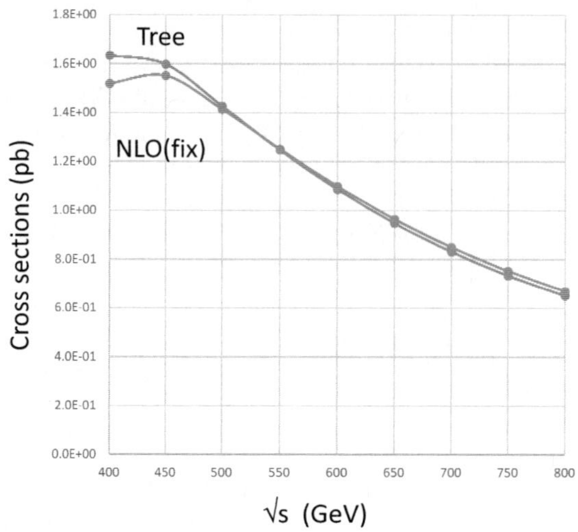

Figure 1. Total cross sections of top quark pair production at tree and one-loop are shown as a function of the CM energy. The blue and orange lines shows the tree and NLO cross sections. The higher order resummation effect is not included.

the weak correction degree is defined as

$$\delta_{weak} = \frac{\sigma_{NLO;ISR} - \sigma_{ISR}}{\sigma_{ISR}}. \tag{7}$$

In the definition of δ_{weak}, the trivial initial state photonic corrections are subtracted from the full $\mathcal{O}(\alpha)$ electroweak radiative corrections. Thus, we can understand δ_{weak} shows a fraction of the pure-weak correction in the full $\mathcal{O}(\alpha)$ electroweak radiative corrections. The behaviors of δ_{weak} and δ_{NLO} are shown in Fig. 3 with respect to the CM energies. One can see the pure-weak correction becomes smaller and smaller at a high energy region, and arrive at almost zero at the CM energy of 800 GeV. On the other hand, at the CM energy of 500 GeV, the pure-weak corrections gives +4% correction over the trivial ISR corrections.

3.3. Photonic correction at two-loop order

As mentioned above, the structure function $H(s, x)$ are known up to two-loop order. All of above results are obtained using formulae including two-loop contributions. As also mentioned in previous subsection, a main contribution of the radiative corrections comes from the initial state photonic correction at the high energy region. Therefore, the ISR correction is important among many terms of the radiative corrections. If the two-loop contribution has a large fraction over the one-loop ones, one need to consider even higher-loop corrections for the future experiments. The fraction of two-loop contribution

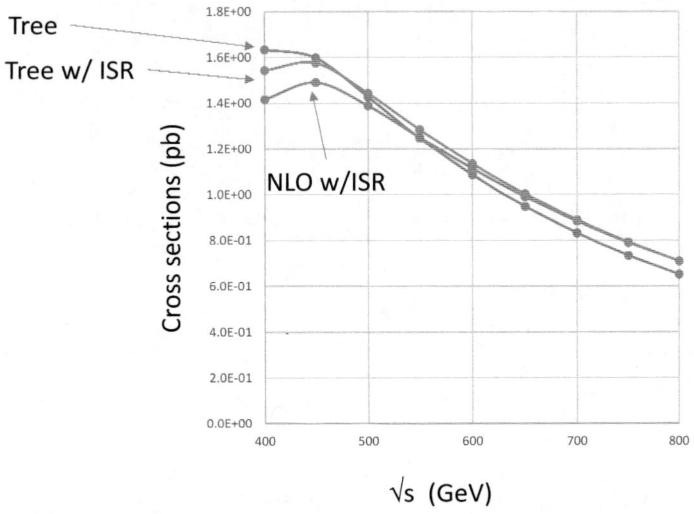

Figure 2. Total cross sections of top quark pair production at the tree, tree with the ISR and one-loop with the ISR corrections are shown as a function of the CM energy. The blue and orange lines shows the tree and NLO cross sections with the ISR. The gray line shows the cross sections with only the ISR correction on the tree cross sections.

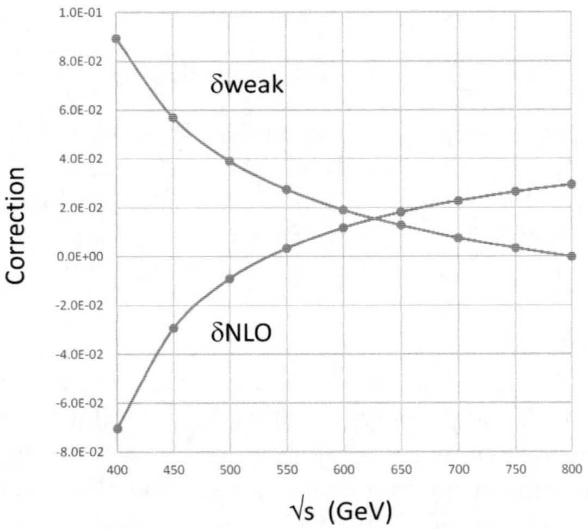

Figure 3. The correction degree of the full $\mathcal{O}(\alpha)$ electroweak radiative corrections (blue line) and the pure-weak corrections (orange line).

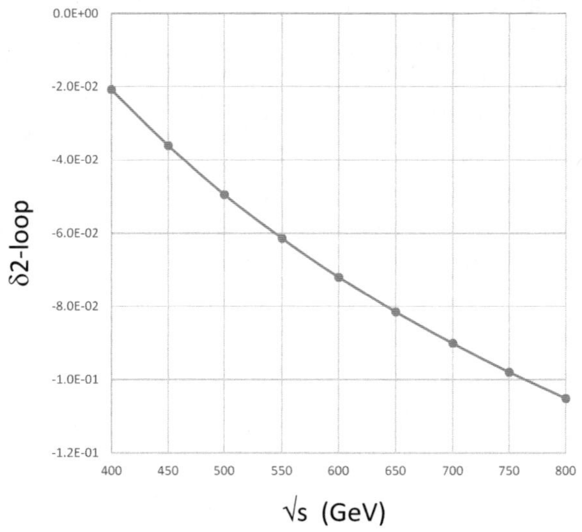

Figure 4. The contribution of the two-loop diagrams on the ISR corrections are show as a function of the CM energies. The definition of $\delta_{2-\text{loop}}$ is given in (8).

over the one-loop one is defined as

$$\delta_{2-\text{loop}} \quad = \quad \frac{\sigma_{ISR}(2-\text{loop}) - \sigma_{\text{ISR}}(1-\text{loop})}{\sigma_{ISR}(1-\text{loop})}, \tag{8}$$

where $\sigma_{ISR}(1-\text{loop})$ and $\sigma_{ISR}(2-\text{loop})$ show the ISR corrected cross sections by the structure function with one-loop diagrams only and with both one- and two-loop diagrams, respectively. One can see in Fig. 4 that the two-loop contribution becomes larger according to increase the CM energy. It is still about -5% at the CM energy of 500 GeV and reaches about -10% at the CM energy of 800 GeV.

4. Summary

We calculated the precise cross sections of an $e^+e^- \to t\bar{t}$ process at an energy region from 400 GeV to 800 GeV. Especially, the initial-state photon emissions are discussed in details. A resummation technique is applied for the initial-state photon emission up to two-loop order. We found that the total cross section of a top quark pair-production at a center of mass energy of 500 GeV receives the pure-weak corrections of $+4\%$ over the trivial ISR corrections. Among the ISR contributions, two-loop diagrams gives -5% correction with respect to the one-loop ones at the CM energy of 500 GeV.

References

[1] G. Aad, et al., Observation of a new particle in the search for the Standard Model Higgs boson with the ATLAS detector at the LHC, Phys. Lett. B716 (2012) 1–29. `doi:10.1016/j.physletb.2012.08.020`.

[2] S. Chatrchyan, et al., Observation of a new boson at a mass of 125 GeV with the CMS experiment at the LHC, Phys. Lett. B716 (2012) 30–61. doi:10.1016/j.physletb.2012.08.021.

[3] T. Behnke, J. E. Brau, B. Foster, J. Fuster, M. Harrison, J. M. Paterson, M. Peskin, M. Stanitzki, N. Walker, H. Yamamoto, The International Linear Collider Technical Design Report - Volume 1: Executive SummaryarXiv:1306.6327.

[4] H. Baer, T. Barklow, K. Fujii, Y. Gao, A. Hoang, S. Kanemura, J. List, H. E. Logan, A. Nomerotski, M. Perelstein, et al., The International Linear Collider Technical Design Report - Volume 2: PhysicsarXiv:1306.6352.

[5] J. Fujimoto, Y. Shimizu, Radiative Corrections to $e^+e^- \to t\bar{t}$ in Electroweak Theory, Mod. Phys. Lett. 3A (1988) 581. doi:10.1142/S0217732388000696.

[6] J. Fleischer, T. Hahn, W. Hollik, T. Riemann, C. Schappacher, A. Werthenbach, Complete electroweak one loop radiative corrections to top pair production at TESLA: A ComparisonarXiv:hep-ph/0202109.

[7] J. Fleischer, A. Leike, T. Riemann, A. Werthenbach, Electroweak one-loop corrections for e^+e^--annihilation into $t\bar{t}$ including hard bremsstrahlung, The European Physical Journal C - Particles and Fields 31 (1) (2003) 37–56. doi:10.1140/epjc/s2003-01263-8.

[8] P. H. Khiem, J. Fujimoto, T. Ishikawa, T. Kaneko, K. Kato, Y. Kurihara, Y. Shimizu, T. Ueda, J. A. M. Vermaseren, Y. Yasui, Full o(α) electroweak radiative corrections to $e^+e^- \to t\bar{t}\gamma$ with grace-loop, The European Physical Journal C 73 (4) (2013) 2400. doi:10.1140/epjc/s10052-013-2400-3.

[9] N. M. U. Quach, Y. Kurihara, K. H. Phan, T. Ueda, Beam polarization effects on top-pair production at the ILCarXiv:1706.03432.

[10] G. Bélanger, F. Boudjema, J. Fujimoto, T. Ishikawa, T. Kaneko, K. Kato, Y. Shimizu, Automatic calculations in high energy physics and grace at one-loop, Physics Reports 430 (3) (2006) 117 – 209. doi:http://dx.doi.org/10.1016/j.physrep.2006.02.001.

[11] J. Fujimoto, T. Ishikawa, Y. Kurihara, M. Jimbo, T. Kon, M. Kuroda, Two-body and three-body decays of charginos in one-loop order in the mssm, Phys. Rev. D 75 (2007) 113002. doi:10.1103/PhysRevD.75.113002.

[12] G. Bélanger, F. Boudjema, J. Fujimoto, T. Ishikawa, T. Kaneko, Y. Kurihara, K. Kato, Y. Shimizu, Full $\mathcal{O}(\alpha)$ electroweak corrections to double higgs-strahlung at the linear collider, Physics Letters B 576 (2003) 152 – 164. doi:http://dx.doi.org/10.1016/j.physletb.2003.09.080.

[13] G. Bélanger, F. Boudjema, J. Fujimoto, T. Ishikawa, T. Kaneko, K. Kato, Y. Shimizu, Y. Yasui, Full $\mathcal{O}(\alpha)$ electroweak and $\mathcal{O}(\alpha_s)$ corrections to $e^+e^- \to t\bar{t}h$, Physics Letters B 571 (2003) 163 – 172. doi:http://dx.doi.org/10.1016/j.physletb.2003.07.072.

[14] G. Bélanger, F. Boudjema, J. Fujimoto, T. Ishikawa, T. Kaneko, K. Kato, Y. Shimizu, Full $\mathcal{O}(\alpha)$ corrections to $e^+e^- \to \nu\bar{\nu}h$ by grace, Nuclear Physics B - Proceedings Supplements 116 (2003) 353 – 357, proceedings of the 6th International Symposium on Radiative Corrections and the 6th Zeuthen Workshop on Elementary Particle Theory. doi:http://dx.doi.org/10.1016/S0920-5632(03)80198-6.

[15] K. Kato, F. Boudjema, J. Fujimoto, T. Ishikawa, T. Kaneko, Y. Kurihara, Y. Shimizu, Y. Yasui, Radiative corrections for Higgs study at the ILC, PoS HEP2005 (2006) 312.

[16] K.-i. Aoki, Z. Hioki, R. Kawabe, M. Konuma, T. Muta, Electroweak theoryframework of on-shell renormalization and study of higher-order effects, Progress of Theoretical Physics Supplement 73 (1982) 1. doi:10.1143/PTPS.73.1.

[17] J. Fujimoto, M. Igarashi, N. Nobuya, S. Yoshimitsu, T. Keijiro, Radiative corrections to e+e reactions in electroweak theory, Progress of Theoretical Physics Supplement 100 (1990) 1. doi:10.1143/PTPS.100.1.

[18] J. A. M. Vermaseren, New features of FORM, For the newest verion of FORM, see [26]. arXiv:math-ph/0010025.

[19] G. J. van Oldenborgh, FF: A Package to evaluate one loop Feynman diagrams, Comput. Phys. Commun. 66 (1991) 1–15. doi:10.1016/0010-4655(91)90002-3.

[20] T. Hahn, M. Perez-Victoria, Automatized one loop calculations in four-dimensions and D-dimensions, Comput. Phys. Commun. 118 (1999) 153–165. doi:10.1016/S0010-4655(98)

00173-8.

[21] S. Kawabata, A new monte carlo event generator for high energy physics, Computer Physics Communications 41 (1) (1986) 127 – 153. doi:http://dx.doi.org/10.1016/0010-4655(86)90025-1.

[22] S. Kawabata, A new version of the multi-dimensional integration and event generation package bases/spring, Computer Physics Communications 88 (2) (1995) 309 – 326. doi:http://dx.doi.org/10.1016/0010-4655(95)00028-E.

[23] F. Boudjema, E. Chopin, Double Higgs production at the linear colliders and the probing of the Higgs selfcoupling, Z. Phys. C73 (1996) 85–110. doi:10.1007/s002880050298.

[24] P. Khiem, Y. Kurihara, J. Fujimoto, T. Ishikawa, T. Kaneko, K. Kato, N. Nakazawa, Y. Shimizu, T. Ueda, J. Vermaseren, Y. Yasui, Full electroweak radiative corrections to at the ilc with grace-loop, Physics Letters B 740 (2015) 192 – 198. doi:http://dx.doi.org/10.1016/j.physletb.2014.11.048.

[25] C. Patrignani, et al., Review of Particle Physics, Chin. Phys. C40 (10) (2016) 100001. doi:10.1088/1674-1137/40/10/100001.

[26] [link].
URL https://www.nikhef.nl/~form/

Institute of Physics
Dirac House, Temple Back
Bristol BS1 6BE UK

ISSN: 1742-6588
ISBN 978-1-5108-5213-6

9 781510 852136

CURRAN ASSOCIATES INC.
proceedings.com